Asian Tsunami and Social Work Practice: Recovery and Rebuilding

Asian Tsunami and Social Work Practice: Recovery and Rebuilding has been co-published simultaneously as *Journal of Social Work in Disability & Rehabilitation*, Volume 5, Numbers 3/4 2006.

Asian Tsunami and Social Work Practice: Recovery and Rebuilding

Ngoh Tiong Tan, PhD
Allison Rowlands, PhD
Francis K. O. Yuen, DSW
Editors

Asian Tsunami and Social Work Practice: Recovery and Rebuilding has been co-published simultaneously as *Journal of Social Work in Disability & Rehabilitation*, Volume 5, Numbers 3/4 2006.

Routledge
Taylor & Francis Grou
NEW YORK AND LONDO

First published by
The Haworth Press, Inc.
10 Alice Street
Binghamton, N Y 13904-1580

This edition published 2011 by Routledge

Routledge
Taylor & Francis Group
711 Third Avenue
New York, NY 10017

Routledge
Taylor & Francis Group
2 Park Square, Milton Park
Abingdon, Oxon OX14 4RN

Asian Tsunami and Social Work Practice: Recovery and Rebuilding has been co-published simultaneously as *Journal of Social Work in Disability & Rehabilitation*™, Volume 5, Numbers 3/4 2006.

Cover design by Jennifer Gaska.

Library of Congress Cataloging-in-Publication Data

Asian Tsunami and Social Work Practice: Recovery and Rebuilding/Ngoh Tiong Tan, Allison Rowlands, Francis K. O. Yuen, editors.

p. cm.

"Co-published simultaneously as Journal of Social Work in Disability & Rehabilitation, Volume 5, Numbers 3/4 2006"

Includes bibliographical references and index.

ISBN 13: 978-0-7890-3235-5 (hard cover : alk. paper)

ISBN 10: 0-7890-3235-X (hard cover : alk. paper)

ISBN 13: 978-0-7890-3236-2 (soft cover : alk. paper)

ISBN 10: 0-7890-3236-8 (soft cover : alk. paper)

1. Disaster relief–Indian Ocean Region. 2. Indian Ocean Tsunami, 2004. 3. Social service–Indian Ocean Region. 4. Disaster victims–Services for–Indian Ocean Region. I. Tan, Ngoh Tiong. II. Rowlands, Allison. III. Yuen, Francis K. O. IV. Journal of social work in disability & rehabilitation.

HV555.I35 A75 2006

363.34'94532091824–dc22

2006025366

This publication is dedicated to the survivors of the tsunami, especially the children and families, for their courage in rebuilding their lives and communities as well as to the many unsung heroes, volunteers and professionals, who served well the victims of the tsunami.

Asian Tsunami and Social Work Practice: Recovery and Rebuilding

CONTENTS

ABOUT THE EDITORS

Ngoh Tiong Tan, PhD, is Professor of Social Work at Augsburg College, Minneapolis, MN. He was Chair of FAST (Families And Survivors of Tsunami) Project and co-chair of Commonwealth Organisation for Social Work. He was Associate Professor and former Vice Dean at National University of Singapore and former Regional President, International Federation of Social Workers (Asia Pacific). Dr. Tan is active in international social work and has been consultant to different social work programs in Vietnam, Sri Lanka, Myanmar and China. Dr. Tan's research has been on conflict, family, social support, social work models and social development and he has published and edited several books including: three volumes of *Social Work Around the World, Social Care in Asia, Human Rights Perspective, Alternative Dispute Resolution, People Effectiveness, Personality and Effective Management, Extending Frontiers: Social Work and Social Issues in Singapore*. He is also Associate Editor of *Asia Pacific Journal of Social Work and Development*. His numerous articles have appeared in *Social Development Issues, Families In Society, Journal of Youth Studies, Mediation Quarterly, Family and Conciliation Courts Review, Journal of Conflict Resolution* and *Ageing International.*

Allison Rowlands, PhD, currently manages the social work service in Centrelink, Area Hunter, Australia. She was Secretary of FAST (Families And Survivors of Tsunami) Project and was Senior Lecturer, Department of Social Work at the National University of Singapore. Her disaster recovery experience includes community recovery coordination, policy development and training in Australia, as well as training and consultation in psychosocial rehabilitation in Timor Leste. Her research interests are in disaster recovery, and rehabilitation and disability, particularly acquired brain injury. She has published in *Families in Society, British Journal of Therapy and Rehabilitation, Brain Impairment, Australian Child and Family Welfare*, and *Asia Pacific Journal of Social Work and Development*, as well as co-editing *Social Work Around the World III*. Prior to taking up appointment at NUS, Dr. Rowlands was a Business Manager in Centrelink, and Lecturer in Social Work at the University of Newcastle, Australia.

Francis K. O. Yuen, DSW, ACSW, is Professor in the Division of Social Work at California State University, Sacramento. He is widely published in the areas of family health social work practice, disability, grant writing, program evaluation, substance abuse, international social work, and services to the refugees and immigrants. He is a member of several editorial boards, including that of the *Journal of Ethnic & Cultural Diversity in Social Work*. He has authored and edited many books, including several for The Haworth Press, Inc., such as *Social Work Practice with Children and Families: A Family Health Approach* (2005), *International Perspective on Disability Services: The Same But Different* (2004), and the upcoming *Disability and Social Work Education: Practice and Policy Issues* (2007). He is also a co-editor for *Handbook on Emotional and Behavioural Difficulties* (2004) and a co-author for *Practical Grant Writing and Program Evaluation* (2003).

About the Contributors

Pattamaporn Busaspathumrong, PhD, is currently the Program Leader in Humanities and Social Sciences, Faculty of Liberal Arts, Asian University, Chon Buri, Thailand. She received her Doctoral degree in Social Work from the University of Pennsylvania, USA. Her education in the US was supported by the King Anandha Mahidol Scholarship. Her experience includes work in a Child Advocacy Unit and in the Defenders' Association of Philadelphia, and consultancies to the United Nations and UNICEF. Her current research area is welfare mix.

Laura Hemingway, MA, is a doctoral research student at the Centre for Disability Studies, University of Leeds, UK. Her past and current research interests include the situation of disabled people in post-conflict areas and the barriers to home ownership and finance for disabled people in Britain.

Tay Keong Tan, PhD, is Executive Director of the Singapore International Foundation, an agency involved in Tsunami reconstruction in Sri Lanka and Indonesia. He serves as a member of the Tsunami Reconstruction Facilitation Committee responsible for coordinating the involvement of Singaporean non-governmental organisations in the reconstruction efforts. Dr. Tan has a Master's and a PhD in Public Policy from Harvard University. He previously taught public ethics and public management at the Lee Kuan Yew School of Public Policy at the National University of Singapore. He also served two years at United Nations Headquarters in New York as the Special Assistant to the Under-Secretary-General and Chief of Staff of the Office of Internal Oversight Services.

Caroline Manning, BSW (Hons), is Business Manager, Social Work Services, in Centrelink's National Office, Canberra, Australia. Centrelink is an Australian Government agency delivering a range of services on behalf of 25 purchasing departments, organisations and other agencies to the Australian community. Services include the provision of income support payments, employment and training assistance, and assistance to people with difficult personal and family issues.

Stewart Millar, BA (Social Work) is Senior Social Worker at Launceston Call Centre, Centrelink, in Launceston, Australia.

Vijay K. Nagaraj, MSW, has worked as a political and human rights activist and presently is a faculty member at the Tata Institute of Social Sciences, Mumbai, India. His research interests include social movements and politics, human rights, public policy accountability and impunity, involuntary resettlement and rehabilitation.

Tania Newton, BSW, is Business Manager, Social Work, Area North Australia, Centrelink, in Darwin, Australia.

Mark Priestley, PhD, is Reader in Disability Studies at the Centre for Disability Studies, University of Leeds, UK and administrator of the international discussion group, disability-research@jiscmail.ac.uk. His current research interests include disability policies in the UK and in comparative and European context. He teaches courses in disability studies and research methods and has been a member of the editorial boards of three international journals: *Disability & Society, Handicap: revue de sciences humaines et socials*, and the *Scandinavian Journal of Disability Research*.

Beverley Raphael, AM, MBBS, MD, FRANZCP, FRCPsych, FASSA, Hon MD, is Emeritus Professor of Psychiatry, University of Queensland, Professor of Population Mental Health & Disasters at the University of Western Sydney, and Professor of Psychological Medicine at the Australian National University. She is known internationally for her expertise in trauma, grief and disasters, having published and researched widely in this field. She has been involved in disaster response to a broad range of natural and man-made disasters, including response to the Bali bombing and the Indian Ocean Tsunami. Her long-term interests include prevention in mental health, women's mental health, family issues and issues affecting consumers in mental health care systems. She has also been involved in the development of Aboriginal Mental Health Policy in Australia.

Garry Stevens, BSc (Hons), M Clin Psych, MAPS, is a Senior Clinical Psychologist with extensive experience in the field of child and adolescent mental health. He has previously researched and developed brief risk assessment and therapy models in child and adolescent clinical services. He has been involved more recently in the field of disaster mental health, including the response to the Indian Ocean Tsunami and the Bali Bombing. He is current Senior Research Fellow at the MH-DAT Unit (Mental Health aspects of Disaster and Terrorism), University of Western Sydney, where he coordinates research programmes.

Krithika Srinivasan, MSW, is Programme Associate at Tata Institute of Social Sciences, Mumbai, India. Her research interests include environmental governance and policy, urban habitats and disasters.

Paige Johnson Tan, PhD, completed her doctoral studies in Foreign Affairs, specializing in Comparative Politics and Southeast Asia, at the University of Virginia. Her academic specialty is the role of political parties in democratization, with special emphasis on Indonesia. Dr. Tan, now Assistant Professor at the University of North Carolina, Wilmington in the United States, has contributed articles on politics to *Contemporary Southeast Asia* (Singapore), *Education about Asia* (US), the *Jakarta Post* (Indonesia), the *Van Zorge Report on Indonesia* (Indonesia), and the *Daily Yomiuri* (Japan). Dr. Tan is on leave from her university this academic year carrying out research in Southeast Asia.

Sharon Webb, BSW, is Business Manager, Social Work, Area South Metro, Centrelink, in Liverpool, Australia.

Preface: Contribution of Social Work to Recovery and Rehabilitation After the Asian Tsunami Disaster

The Asian Tsunami of December 2004, Hurricane Katrina which hit the Gulf Coast of the United States in August 2005, and the earthquake that struck the remote regions in Pakistan in October 2005 wreaked havoc on hundreds of thousands of lives. Many have lost their loved ones and homes. The survivors, especially the women and children, are most vulnerable and at risk. Local and international rescue and recovery have contributed towards the process of rehabilitation.

Natural disasters such as earthquakes, floods and fire, and disasters attributable to human action or intent, such as environmental pollution, nuclear accidents, airplane crashes and wars–all these cause death, injury and disruption to life. Well-planned and timely measures need to be taken to find new levels of equilibrium. The rehabilitation process often requires resources beyond the immediate intervention to long-term strategies involving empowerment and capacity building.

Much of the aid in the initial aftermath phase of the Tsunami focused on the physical needs in terms of food, water, shelter and medical relief. The longer-term need is the social and emotional support to survivors and the recovery and rebuilding of family and community after the traumatic experience.

When disaster strikes the impact is often not just on individuals but also on entire communities, nations, and the world. This was also evi-

---

[Haworth co-indexing entry note]: "Preface: Contribution of Social Work to Recovery and Rehabilitation After the Asian Tsunami Disaster." Rowlands, Allison, Ngoh Tiong Tan, and Francis K. O. Yuen. Co-published simultaneously in *Journal of Social Work in Disability & Rehabilitation* (The Haworth Press, Inc.) Vol. 5, No. 3/4, 2006, pp. xxv-xxxi; and: *Asian Tsunami and Social Work Practice: Recovery and Rebuilding* (ed: Ngoh Tiong Tan, Allison Rowlands, and Francis K. O. Yuen) The Haworth Press, Inc., 2006, pp. xix-xxv. Single or multiple copies of this article are available for a fee from The Haworth Document Delivery Service [1-800-HAWORTH, 9:00 a.m. - 5:00 p.m. (EST). E-mail address: docdelivery@haworthpress.com].

Available online at http://jswdr.haworthpress.com

dent in the 9/11 incident. Social workers from all over the world, using their professional capacity, are interested in contributing to relief efforts, especially in the longer-term rebuilding of families and communities.

SOCIAL WORK PERSPECTIVE

Social work seeks to enhance community expertise and empowerment in the decision-making process through all stages of planning, intervention and recovery. The strengths and resilience of the local community are acknowledged. International assistance should be collaborative with national or local partners in a consultancy role and in training or support, such as research and evaluation.

Social workers have much to contribute through the knowledge, skills, and value base of the profession. Social workers can intervene at the individual, group and family, community and policy levels. The intervention, such as training and consultation, should be guided by values and codes of ethics and also disaster recovery principles, such as those of the Sphere Project (Sphere Project, 2004).

We know from our own knowledge base and the literature on disaster recovery management that coming to terms with trauma and disaster is a challenging process which everyone goes through to a greater or lesser degree. People will recover through retracing their past and reevaluating their future goals, whether that is in private, with their family, friends and neighbours, or with professional helpers (van Ommeren, Saxena & Saraceno, 2005). In addition to individual psychosocial help, community interventions can facilitate understanding of what has happened and provide opportunities for working through the traumatic experience and provide support to do so. Encouraging active community involvement in the recovery process is an internationally accepted principle in disaster recovery management (Emergency Management Australia, 1996; Sphere Project, 2004). Community recovery after disaster is therefore informed by community development principles, underscoring the relevance of this mode of intervention from within social work to the interagency effort to support individuals and communities rebuild after disaster (Emergency Management Australia, 2003; Ife, 2002). Recognition of both suffering and strengths is a critical balance to strike in planning for and implementing community recovery interventions (Raphael, 1986).

This special volume, co-sponsored by the International Federation of Social Workers (IFSW), gathered information on projects and ideas from social workers and relief missions to different countries, as well as reviews of the state of research on community recovery after disaster. We are particularly keen to explore the specific social interventions with regard to disability and rehabilitation issues to assist the communities affected by the Asian Tsunami. This includes how social work and other professionals in the affected areas as well as the international community are responding to the disaster, and the impact of their collaborative efforts.

International Social Work Response

The International Federation of Social Workers (IFSW) links some half a million social workers in 80 countries. This global organization strives for social justice and social development through international cooperation between professional organizations (IFSW, 2006). Responding to the crisis, Tom Johannesen, Secretary General of IFSW, said in supporting regionally-based initiatives, that he "looks towards social work-based rebuilding and focusing on children and families," and that "we will assist in mobilizing human and financial resources" (FAST Project, 2006, p. 1). Immediately after that, in January 2005, IFSW issued a call for social workers to assist, especially with the psychosocial response to the Tsunami disaster.

In Japan, Kozo Iwasaki, Treasurer, ISFW Asia-Pacific wrote, "... as social workers, we will not lose hope. We believe that we can help to rebuild lives and families" (FAST Project, 2006, p. 1). Indeed hope, when in the face of despair, is vital in the recovery process. Japanese, Australian and New Zealand social workers and students mounted projects to raise funds for tsunami affected countries. Many social workers and volunteers from the nations of the world, including Singapore, Sweden and the United Kingdom, were mobilized to assist in the recovery process.

In the United States, the National Association of Social Workers and the Council on Social Work Education also issued a statement about the tsunami, mourning "the loss of life and devastation suffered by the people of Bangladesh, India, Indonesia, and Sri Lanka as a result of the massive tsunamis following the earthquake off the coast of Northern Sumatra on December 26. Countries as far away as Kenya, Tanzania, and Somalia in East Africa also were directly affected by these tsunamis" (CSWE, 2006, p. 1). The statement recorded that, "Recovery and reconstruction efforts are underway supported by the spontaneous response to this tragedy by governments, organizations, and individuals. We urge all social workers wishing to contribute to the recovery efforts to follow the advice provided by experts and not send clothes, materials, or any items needing to be transported. We also advise our members that financial contributions will be needed not just for the immediate needs but to assist with the long-term reconstruction" (CSWE, 2006, p. 1).

In relation to affected nations in the south Asian region, Bambang Laksmono, former President of the Indonesian Association of Professional Social Workers (IPSPI), wrote that it was "a very sad period in Indonesia . . . Our country has lost over 100,000 people . . . will take a very long time to heal and we need the help of the international community. We are grateful for the many relief efforts that are taking place. We know that while this is good, we need long-term support to help the families and victims of the Tsunami to come to terms with the disaster. They need to find meaning in their lives and to continue to live despite the tragic loss of their loved ones, livelihood and home" (FAST Project, 2006, p. 1). Indonesian social workers were deployed in Aceh and the social work association mounted training for recovery workers from both government and NGO sectors.

The President of Sri Lankan Association of Professional Social Workers commented, "Here in Sri Lanka, we are grappling with the major devastation that the tsunami has inflicted on our people. Over 30,000 people have lost their lives and many of our children have been orphaned in one fell swoop. This is not the way children should be brought up, without their parents. This disaster will definitely bring about a lot of future issues. Poverty, a feeling of helplessness, Post Traumatic Stress Disorder–all these are very real issues. We are relieved that the world community of social workers is working together to help with this period" (FAST Project, 2006, p. 1). Students and social workers in Sri Lanka developed a demonstration project for psychosocial intervention in the tsunami affected area.

IFSW (Asia and Pacific region) and the Commonwealth Association for Social Work (COSW), together with counterparts in the affected areas, mounted a short-term catalytic project entitled FAST (Family and Survivors of Tsunami). It is heartening to see many Asian social workers and mental health professionals rallying together to work with the survivors and their communities. The FAST project itself is an example of how social workers can use technology to build and organize community.

The FAST project funded work which includes training of frontline workers, demonstration projects and intervention aimed at community rebuilding and education. Through these projects FAST proposes to meet current needs such as the social and emotional rehabilitation and the need for capacity building and development, including community rebuilding. Principles include the management of culturally sensitive support systems, particularly looking into the massive external inputs to relatively culturally isolated communities.

PROFESSIONAL CONTRIBUTION TO DISASTER RELIEF AND REHABILITATION

Whether it is crisis or disaster of tsunami, hurricane, earthquake, avian flu, or SARS–it is vital to be prepared with training and planning. Planning and strategic mobilisation, preparation, and training of workers, enhance the effectiveness of intervention when needed. Clear protocols for systematic actions by the different service providers are expected to yield more efficient operationalization of plans.

It is observed that poorer countries are more exposed to emergencies and crises and there is limited capacity of response agencies to mitigate negative impacts of such events. In this issue, Beverley Raphael and Garry Stevens maintain that what constitutes good public mental health practices for emergency response cuts across diverse cultural settings and that intervention frameworks may be adapted to allow health and recovery workers in different countries to support healing, rehabilitation and harnessing of agency resources.

Mark Priestley and Laura Hemingway's article, "Disability and Disaster Recovery: A Tale of Two Cities?" provides the connections between disability and disaster. Concepts from individual and social models of disability were applied to recovery from the Asian tsunami and Hurricane Katrina. Using the global connection the authors suggest that more attention should be paid to social model approaches such as the linking of poverty and resources for disaster recovery and preparedness.

The article by Ngoh Tiong Tan provides an overview of the plight of the "tsunami generation." Children are most vulnerable and need protection and clear and sustained support for rehabilitation. Policy and programmes should build on people's resilience and provide the psychosocial and community support needed.

Krithika Srinivasan and Vijay Nagaraj's paper provides an analysis of responses to the tsunami in India and identifies factors influencing the outcomes of these responses. The article is particularly useful for understanding the experiences and insights of people affected by the disaster. The need for reliable records as a resource base for informing relief and rehabilitation interventions is discussed in this article.

The article by Caroline Manning, Stewart Millar, Tania Newton and Sharon Webb captures the realities of a unique social work intervention for Australian citizens caught up in the Tsunami in Thailand and Sri Lanka. A range of strategies was provided by these workers and the authors discuss frankly some of the pressures and dilemmas imposed by working for a temporary and intense period in another country and culture.

Tay Keong Tan and Paige Johnson Tan advocate a balance between the need for a quick response to disaster on the one hand and thorough planning on the other. They claim that there is "no right way to provide volunteers or donations for disaster relief," and while having a "Humanitarian Charter and Minimum Standards in Disaster Response" is good, the reality of adhering strictly to the standards and accompanying indicators, is somewhat difficult in practice. The need, thus, is to learn from experience and apply "best practice" as far as possible.

The workshops mounted by IPSPI as part of the FAST Project as well as those by the Singapore Association of Social Workers were directed at equipping participants with the theoretical base and skills to respond effectively at the individual, family and community level after a disaster. Objectives of the training included understanding the psychosocial issues and needs for children and families after disaster, developing core skills in crisis intervention and responding to loss as well as the capacity to undertake a range of assessments after disaster.

Allison Rowlands' article highlights the importance of training for disaster recovery interventions. These programs will vary according to the needs of the participants. For example, qualified professionals may have knowledge, values and a skills base to draw on for their practice; volunteers and other recovery workers would require basic know-how on human responses to disaster and basic intervention techniques. The degree of preparedness of workers will have a bearing on the recovery of communities affected by disaster.

Pattamaporn Busaspathumrong's report deals with the role of social workers and social service delivery for survivors of the tsunami in Thailand. This descriptive paper provides a better understanding of social workers' involvement with more specific practice situations. She advocates for greater collaborations between local, national and international organizations as well as between social workers and other professionals in crisis intervention.

For most of us in our daily practice, the Asian Tsunami has already retreated from the forefront of our consciousness, just as the waves themselves subsided from view after causing their havoc and destruction. Yet the time remains salient to reflect on the interventions undertaken and the lessons learned in this past eighteen months or so. We hope that readers of this special collection will find the accounts of practice and the reviews of literature valuable. We further hope that through this publication, the wider professional helping audience will enhance disaster recovery responses in the future, thus making real the promise that rebuilding can restore a community to a more just society than before.

Allison Rowlands
Ngoh Tiong Tan
Francis K. O. Yuen
Editors

REFERENCES

CSWE (2006) *NASW and CSWE Joint Statement About the Victims of the Tsunamis.* Retrieved on 12 March, 2006: http://www.cswe.org/Katrina/

Emergency Management Australia (1996) *Australian Emergency Manual Disaster Recovery.* Canberra: Emergency Management Australia.

Emergency Management Australia (2003) *Community Development in Recovery from Disaster Part III.* Canberra: Emergency Management Australia.

FAST Project (2006) *Message from IFSW.* Retrieved on 12 March, 2006: http://www.FASTproject.net

Ife, J. (2002, 2nd ed.) *Community Development: Community-Based Alternatives in an Age of Globalisation.* Frenchs Forest Australia: Pearson.

IFSW (2006) *Introduction to IFSW.* Retrieved on 12 March, 2006: http://www.ifsw.org/en/p38000002.html

Raphael, B. (1986) *When Disaster Strikes.* New York: Basic Books.

Sphere Project (2004) *Humanitarian Charter and Minimum Standards in Disaster Response.* Geneva: Sphere Project. Retrieved on 22 December, 2005: http://www.sphereproject.org/handbook/index.html

van Ommeren, M., Saxena, S. & Saraceno, B. (2005) Mental and social health during and after acute emergencies: Emerging consensus? *Bulletin of the World Health Organization*, 83 (1) 71-74.

Disaster and Response: Science, Systems and Realities

Beverley Raphael
Garry Stevens

SUMMARY. The Indian Ocean Tsunami of December 2004 was one of the most significant natural disasters in modern history. The response that it drew was unprecedented at both local and international levels. While a range of specific impacts and losses may be predicted within the affected populations, there is substantial evidence that adverse impacts upon mental health may represent one of the most significant outcomes of such events. People from poorer countries are disproportionately exposed to such emergencies and may experience greater psychosocial burden in the aftermath, as well as a range of adaptations. This raises important questions regarding the current capacity of response agencies to mitigate negative impacts but also concerning the cultural context in which such mental health effects are defined. It also calls into question whether "indicated" interventions are both realistic and ecologically valid in non-Western settings. The science of Disaster Mental Health is a

Beverley Raphael, MD, is Emeritus Professor of Psychiatry and Professor of Population Mental Health & Disasters, University of Western Sydney, Medical School Project Office, 1st Floor, Building EZ, Parramatta Campus, Locked Bag 1797, Penrith South NSW DC 1797, Australia.

Garry Stevens, M Clin Psych, is Senior Research Fellow, University of Western Sydney, Medical School Project Office, 1st Floor, Building EZ, Parramatta Campus, Locked Bag 1797, Penrith South NSW DC 1797, Australia.

[Haworth co-indexing entry note]: "Disaster and Response: Science, Systems and Realities." Raphael, Beverley, and Garry Stevens. Co-published simultaneously in *Journal of Social Work in Disability & Rehabilitation* (The Haworth Press, Inc.) Vol. 5, No. 3/4, 2006, pp. 1-22; and: *Asian Tsunami and Social Work Practice: Recovery and Rebuilding* (ed: Ngoh Tiong Tan, Allison Rowlands, and Francis K. O. Yuen) The Haworth Press, Inc., 2006, pp. 1-22. Single or multiple copies of this article are available for a fee from The Haworth Document Delivery Service [1-800-HAWORTH, 9:00 a.m. - 5:00 p.m. (EST). E-mail address: docdelivery@haworthpress.com].

Available online at http://jswdr.haworthpress.com

doi:10.1300/J198v05n03_01

relatively new field, currently engaged in a number of such debates. However, there is some emerging consensus about what constitutes good public mental health practice throughout the phases of emergency response and across diverse cultural settings. Existing data and practice guidelines provide frameworks that may be adapted to allow health and recovery workers to identify and reduce mental health morbidity, support healing and harness the agency of those affected for the recovery process. doi:10.1300/J198v05n03_01 *[Article copies available for a fee from The Haworth Document Delivery Service: 1-800-HAWORTH. E-mail address: <docdelivery@haworthpress.com> Website: <http://www.HaworthPress.com>*

KEYWORDS. Disaster, psychosocial, Tsunami, public health, Asia

By definition, disasters are overwhelming events which threaten or cause injury or death, damage to property, serious disruption to community functioning and are beyond the normal response capacity of emergency services, requiring a special mobilisation of resources. They may be brief or protracted, arrive without warning to unprepared communities or be anticipated due to their cyclical nature, community knowledge or specific early-warning systems. They may be natural events or technological accidents, such as road transport crashes or factory explosions, or acts of terrorism. Existing disaster models that conceptualise the "phases" of an event, e.g., warning, threat, impact, inventory, rescue and recovery (e.g., Raphael & Wilson 1993), provide a useful profile of the physical and psychological dynamics of a single-incident disaster. The disaster response model PPRR (Prevention, Preparedness, Response and Recovery), which is used internationally, is conceptually linked to the management of event phases such as this, although not in strict sequences (Salter, 1995). It has also been argued that modern disaster planning must encompass "complex emergencies"; such as the compounding effects of famine, civil conflict or environmental destruction, onto which natural disaster events may then be superimposed (e.g., Salama, Spiegel, Talley & Waldman, 2004).

Complex disasters more often reflect the experience of developing countries and are associated with the individual and social effects of chronic stress (Spiegel, 2005). These scenarios were observed in Aceh, Indonesia and North Eastern Sri Lanka following the Indian Ocean Tsunami in 2004. The term "slow disaster" has also been used to describe protracted events such as the HIV epidemic, Avian Influenza or climate change, although there have been debates regarding the usefulness of this concept (Quarantelli, 1998).

As will be discussed, the significant mental health impacts of disasters may often indicate the need for targeted interventions but also raise the conceptual issues of when, with whom, what type and to what ends such interventions should be applied. Population health models provide a conceptual and strategic framework with which health and mental health services may seek to prevent or minimise such morbidity. This approach utilises the data from systematic epidemiological studies of incidence, prevalence, risk and protective factors. The Institute of Medicine (Mrazek & Haggerty, 1994) has conceptualised a spectrum of intervention approaches at population level ("Universal"), with vulnerable sub-populations ("Selected") and prodromal groups ("Indicated") as well as at the levels of health system programs (i.e., primary, secondary and tertiary). The aim of this approach is to address health system capacities and create opportunities for maximal benefit in terms of outcomes, relative to the resource-base.

Some conceptual elements of this approach may be seen in the "Public Health" approach to mental health adopted by the World Health Organisation (WHO, 2001a), which has influenced the mental health response in many of the regions affected by the Tsunami and which will be discussed further. From this perspective the "science" of disaster mental health must articulate patterns of effect at both individual and population levels. In relation to disasters, questions of central concern include (1) likely impacts (mental health and other), (2) the impacts of different exposures, (3) potential patterns of reaction, (4) the identification of groups at greater risk and with greatest need, (5) the proposed basis for intervention and (6) the identification of processes of resilience in the populations affected. These determinations allow options for supporting resilience, while also promoting targeted support and interventions for those most at risk.

The stresses experienced by those exposed to disasters will depend on the objective elements of the disaster and also upon individual and communal appraisals of the event, the losses it brings and its substantive timeframe–immediate to longer-term. Impacts may be physical, psychological, community or resource-based or a combination of all of these. Ultimately, many of these impacts will be experienced at an emotional and psychological level. In possibly the broadest review of the disaster literature to date (225 studies), Norris (2005) observed that the most common disaster-specific stressors include bereavement, life threat, injury to self or family members, horror, property damage and financial loss. With regard to the subsequent effects on survivors, adverse impacts upon mental health were found to represent the most prevalent consequences of disasters. In order of frequency, adverse effects included: specific psychological problems (notably PTSD, Major Depression and Generalised Anxiety Disorder), nonspecific distress, physical health problems, chronic problems of living, resource loss and problems specific to youth.

The survivors of the 2004 Tsunami are likely to have experienced multiple intense stressors such as those noted above, although the reporting of specific mental health effects remains limited (Miller, 2005). WHO projections for Tsunami-affected populations indicate that 12 months after the event, up to 4% of the population may have severe and disabling mental health conditions, with a further 20% experiencing mild to moderate impacts (see van Ommeren, Saxena & Saraceno, 2005b). Depression and Non-PTSD anxiety conditions may form the bulk of this group. In Phang-Nga and Krabi provinces, the worst hit regions of southern Thailand, 4700 survivors sought psychiatric help with females (60%) and young adults (20-40 yrs, 36%) being the largest help-seeking groups. One week after the Tsunami, depression (35%) was the most commonly reported mental health state, followed by severe anxiety (28%). However, by the end of the first month these rates had steadily declined to 10% and 7% respectively (Udomratm, 2005). Some reports indicated that rates of PTSD were lower than for other presentations. Nambi (2005) reported that in Tamil Nadu, the worst affected state in India, 75% of those sampled (N = 200) showed some clinical symptoms, which included Depression (21%), Alcohol Abuse (7%), Generalised Anxiety Disorder (3%) and PTSD (0.5%, 1 case only). In another Thai study (reported in Udomratm, 2005) the effects of being displaced from homes and communities were associated with increased rates of PTSD. Eight weeks after the event, 12% of adults displaced from their original communities were experiencing PTSD compared to 7% of those who were not displaced. Children were similarly affected; however, the rates for both groups had reduced significantly at six months. Qualitative studies in India, including interviews and focus groups, found evidence of recurrent emotional and behavioural responses associated with fear, which included sleeplessness, palpitations and poor concentration (reported in Udomratm, 2005).

In the wider disaster literature, there has been debate as to whether particular exposures are likely to induce greater distress, or specific functional or mental health problems. The more comprehensive studies indicate that threats to life (i.e., specific beliefs such as "I thought I was going to die") or injury ("traumatic injury") during disaster are associated with stronger or longer lasting effects on mental health (Maes, Mylle, Delmire & Altamura, 2000). Individuals, who perceive that their life is directly threatened during a disaster, especially when this evokes shock and helplessness, are at increased risk of a range of reactive processes including emotional "numbing," avoidance behaviour, and the reexperiencing of traumatic events. Intense distress or fear may trigger specific psychological and physiological reactions such as dissociation or prolonged elevation in heart rate (e.g., Shalev et al., 1998). There is evidence that these latter reactions are more likely to develop into a pattern of psychiatric morbidity such as Acute Stress Disorder, Post-Traumatic Stress Disorder (PTSD) or other specific conditions such as Depression and Generalised Anxiety Disorder (Birmes et al., 2003). Those who are physically injured during the event are generally found to experience higher rates of PTSD (Hull, Alexander & Klein, 2002). There is evidence, too, that the nature of the threat or loss is often quite specific to outcomes. Survivors of the Mount St. Helen eruption who were bereaved showed high levels of depression, while perceived life threat was strongly associated with the development of PTSD (Shore & Tatum, 1986). Similar patterns have been observed with children exposed to a school sniper (Pynoos & Nader, 1988) and adults in the September 11 attacks (Galea et al., 2003). These studies reflect consistent findings that a "dose-response" exists; whereby increased proximity, duration or frequency of disaster exposure is associated with a worsening of mental health symptoms (see Brewin, Andrews & Valentine, 2000).

In emergency events where there have been mass casualties, exposure to dead bodies may be a profound stressor and has been shown to be associated with increased psychiatric morbidity. Direct survivors, onlookers and rescue workers have reported trauma responses from both the anticipation and experience of witnessing or handling human remains in disaster situations (e.g., Dyregrov, Kristoffersen & Gjestad, 1996). These findings may be particularly relevant to the Tsunami given the massive number of deaths and the high exposure of community members and aid workers in the aftermath. While these associations are not well understood, available research indicates that greater mental health impacts are associated with increased duration of exposure, grotesque or mutilated remains and increased identification or "emotional involvement" with the deceased, especially children (e.g., Ursano, Fullerton & Norwood, 2003).

The location of disaster appears to be a significant factor influencing mental health impacts. Asian countries appear to experience a greater number of natural disasters, possibly due to factors such as weather patterns and population distribution. Norris observed that the Asian region experiences an average of 200 events annually, followed next by the Americas at an average of 111 events (Norris, 2005). In a recent review of natural disasters and mental health in Asia, Kokai, Fuji, Shinfuku, and Edwards (2004) estimated that 3 billion people were affected internationally between 1967 and 1991 but that 85% of these lived in Asia. In addition to frequency, the limited resources in some regions to respond to such emergencies may mean that health and mental health impacts may be greater.

The type of disaster may also strongly influence mental health outcomes. In a previous review, Norris et al. (2002) observed that when other variables are controlled, those exposed to mass violence and terrorism experience severe mental health impairment at almost twice the rate of those exposed to natural or technological disasters (a point that will be discussed further). There was little difference between the latter two categories. It was found, however, that the disaster location (developed vs. developing countries) accounted for 15% of variance in the severity of effects, with the latter being more adversely affected. This represents a substantial influence on outcomes. In fact, the average effects of natural disasters in developing countries were more severe than were the effects of mass violence events in developed countries. Across any type of disaster, people from developing countries experience severe mental health effects at more than twice the rate of those from developed countries (Norris et al., 2002). Given the small number of mental health professionals in many countries, these findings have significant implications for how practical mental health services may be delivered in a number of regions.

DISASTER LOSSES

Frequently, the destructive force, chaos and threat associated with disasters will result in deaths of community members and the loss of social networks and the rituals of everyday life. Survivors often experience a period of disbelief, which may be more pronounced when the event or resulting fatalities are unexpected or involve the loss of children. Shock, denial and a sense that things are "unreal" in the aftermath are commonly reported and may affect entire communities as well as individuals. Key losses may include the deaths of primary attachment figures such as a parent or child, the loss of homes, places of employment and treasured possessions that embodied personal histories or defined important relationships.

Survivors face a number of practical, emotional and psychological challenges in the aftermath. In many regions families who are seeking food, water or shelter, are displaced from their homes or workplaces, or are caring for small children, may have little opportunity to grieve for those who have died, as they must focus on the practical tasks of survival. Some disasters and terrorist attacks may result in prolonged periods of uncertainty about the fate of loved ones. In situations of mass death or disfigurement, identification of the deceased may only be possible through formal Disaster Victim Identification (DVI) processes, such as DNA and dental record analysis. This may not be possible in many circumstances of mass death, such as the Tsunami. Concerns about the spread of disease meant that many of the deceased were buried in mass graves, with those bereaved often not knowing the final resting place of loved ones. Research has indicated that dead bodies in such circumstances do not pose an imminent threat of disease, and WHO has recommended against mass burial as a reflexive practice (Miller, 2005). Access to deceased loved ones and their place of burial, the opportunity to observe religious customs and say "goodbye," may all be factors which can ease suffering and support recovery and resilience.

Bereavement represents the normal reactive process of mourning such losses (Engel, 1961). However, the traumatic circumstances of disaster-related deaths often produce a complex mix of traumatic stress and bereavement phenomenology, which is more properly called a "traumatic bereavement" or "traumatic grief," (Raphael, Martinek & Wooding, 2004). At present, little is known as to whether particularly horrific exposures (such as the mass killings in Rwanda or Bosnia) may produce specific complications beyond traumatic bereavement or PTSD as they are currently conceptualized. However, the wider literature on complicated grief (traumatic and non-traumatic loss) generally indicates a more protracted course for those so affected (greater than six months), more adverse outcomes and greater treatment complexity (Gray, Prigerson & Litz, 2004). Certain groups may be more vulnerable, including those who lose a close family member (i.e., parent, child or spouse), and those who have experienced childhood abuse or early parental loss (e.g., Silverman, Johnson & Prigerson, 2001). For others, prior experiences, skills and other adaptations may act as protective factors against the potential negative effects of trauma and loss. Silove

(1999) has addressed the prolonged and compounding stresses experienced by refugees, including the complex trauma associated with torture and other human rights violations. He has proposed a model regarding the adaptive systems that form the basis of an individual's response to such trauma, and that may support adaptation even in the most extreme circumstances. The model is proposed as a means of integrating research and clinical endeavours with this population. Importantly, it also helps to delineate the respective pathways associated with either ongoing psychiatric disability or psychosocial restitution.

Some studies have indicated that grief and trauma in early life or previously may also be associated with personal growth or "traumatic growth" (e.g., Tedeschi & Calhoun, 2004), such that these individuals may have developed insight and strengths which allow them to cope better with later circumstances of death and loss. Bonnano (2004) cites evidence suggesting that resilience represents a distinct coping style in relation to adversity and that such traits are relatively common. Nonetheless, clinicians working with the protracted effects of traumatic bereavement and some forms of complex trauma, increasingly recognize that traumatic stress effects (i.e., those related to life threat) must often be dealt with first, before it is possible to grieve (Regehr & Sussman, 2004).

Acts of human malevolence, such as terrorism, have been shown to produce higher levels of psychological morbidity than other types of disasters (North, Nixon, Hariat & Mallonee, 1999) and may represent a qualitatively different exposure in psychological terms. The death of loved ones through arbitrary and calculated actions may be difficult for survivors to comprehend and is associated with higher levels of intrusion and avoidance symptoms (Norris, 2005). Such events may also act as a significant and complicating stressor in relation to bereavement (Raphael, Dunsmore & Wooding, 2004). Survivors' preoccupation as to the "meaning" of the death(s), the perpetrators' motivations and thoughts of revenge and justice, may all act as a defence. Ultimately, however, these preoccupations may impede both the acceptance of the loss and its resolution (Raphael & Wooding, 2004).

One of the most tangible aspects of the Tsunami was the massive loss of resources that had occurred and which would later confront those who survived. Resource losses ranged from physical assets such as homes and possessions to valued family and community networks, and the cultural or religious rituals that draw individuals together in common purpose and meaning. The loss of such "social capital" following disasters has been shown to have significant mental health effects

(Buckland & Rahman, 1999). Conversely, specific models developed for building social capital following the earthquakes in Kobe, Japan, were found to improve satisfaction and rates of community recovery following earthquakes in Gujurat, India (Nakagawa & Shaw, 2004). Erikson's (1994) study of a massive flood at Buffalo Creek, West Virginia in 1972 showed that the disruption of neighborhoods and social networks predicted psychological distress as readily as some key personal losses. The effect of being displaced, albeit to safe locations, was also a significant source of stress. Phifer and Norris (1989) showed that the most strongly affected flood survivors were those who experienced both high personal loss and high community destruction, suggesting an interaction of these factors.

The distress associated with resource losses appears to be strongly influenced by the location and economic context in which the losses occur. A cross-national study one month after Hurricane Georges found that distress levels were predicted by location, resource loss and the availability of social support, respectively (Sattler et al., 2002). Distress levels consistent with Acute Stress Disorder were observed in 25% of respondents in Puerto Rico and the Dominican Republic, while only 10% of those in the United States (Alabama, U.S. Virgin Islands) were similarly affected. Rates of depression were also higher in the two former countries. Distress was found to be associated with different types of losses in different regions. "Basic object" loss (food, water, money) was the source of greatest distress in Puerto Rico. "Condition" loss (family and work relationships) accounted for the greatest resource distress in the Dominican Republic, while "object" loss (sentimental items, furniture, etc.) was associated with the greatest distress in the U.S. regions. As will be discussed further, arresting these compounding cycles of loss will form a key aspect of recovery planning, particularly in complex emergency settings and regions with fewer resources.

SYSTEMS: PRE-EMERGENCY PLANNING AND INTERVENTION

The high levels of exposure and mental health burden that may be experienced by those from developing countries raise important questions about how such effects are defined, whether specific interventions and programs can reduce negative effects and how they may be delivered effectively across diverse settings. Epidemiological studies indicate PTSD is both prevalent and disabling in Western societies (Kessler, 2000). It is equally true, however, that most people exposed to trauma do not experience long-lasting mental health effects and that a rapid return to previous functioning occurs for a large majority of people (Kessler, Sonnega, Bromet, Hughes & Nelson, 1995). Post-Traumatic Stress Disorder is probably the single most widely studied mental health condition in post-disaster settings. In the Norris (2005) review, twice as many investigations sought to establish, as their primary index, the prevalence of PTSD (at a ratio of 4:2:1 regarding PTSD, Depression and Anxiety conditions respectively). It has been argued that this preoccupation with PTSD is an example of the imposition of a Western model of illness across cultures, and may promote a pathologising of normal distress (Summerfield, 2005). Scarce resources may be directed towards PTSD "case-finding" and the establishment of separate, specialist trauma-focused treatment services, further fracturing the limited mental health services that exist in many countries (van Ommeren, Saxena & Saraceno, 2005a). Since 2001, WHO have argued that a redirection of mental health policy and practice is needed, so as to support the natural processes of recovery known to occur after emergencies, while at the same time identifying those with significant and potentially longer-term impacts on mental health and functioning. They suggest that this is achievable through an integrated public health approach, incorporating preexisting human and community resources, coordinated social and mental health interventions and broad-spectrum mental health care, including those with preexisting mental health disorders (WHO, 2001a, 2003).

Pre-emergency mental health planning is vital and relates to the establishment of an integrated care strategy for new and emergent mental health presentations after a disaster, while also determining the most effective delivery of available services. Key determinations will include defining those presentations likely to require mental health assessment and care, the available health/mental health services and community resources and their current or potential integration with the primary care system.

Prevalence analysis of conditions such as PTSD in non-Western settings may appear to provide validation of this condition without evidence that it best describes the mental health problems of local survivors (de Jong, Komproe & van Ommeren, 2003). Significant presentations such as depression, somatic complaints, complicated patterns of grief, anxiety and specific "culture bound" conditions, though assumed to be common following the Tsunami and similar events, risk being marginalised through this approach. Debates regarding the public health value of the PTSD construct and the need for vertical (separate) trauma-focused services have caused confusion among program planners. Such debates even led to the omission of a mental health section from early versions of the UN-supported Charter of Minimum Standards for Disaster Response (known as the "Sphere Project," Sphere Project, 2004) due to perceived disagreement among experts (van Ommeren et al., 2005a). These latter authors point to emerging expert consensus on core issues including (1) that exposure to extreme stressors is a risk factor for social and mental health problems and disorders (2) that emergencies may severely disrupt social structures and care systems for those with preexisting disorders and (3) that a range of social, health and mental health interventions are supported, conditional on their being tailored to local contexts, needs and resources.

The World Health Report of 2001 recommends the adoption of a public health perspective encompassing all mental health problems that arise in post-disaster environments, ranging from widespread non-specific psychological distress induced by trauma and loss, to severe preexisting and new mental disorders (WHO, 2001a). The provision of mental health care within primary and general health-care settings is more likely to reduce stigma and enhance access, integration of services, case-finding and longitudinal care. It may also be more pragmatic in developing countries that may have small, fragmented health and mental health workforces and where the former group may be trained in mental health assessment and basic care interventions. Primary health provisions form the central aspect of the eight principles of planning and emergency mental health response proposed by WHO (van Ommeren et al., 2005a). These include (1) contingency planning before the acute emergency, (2) socioculturally sensitive assessment prior to intervention, (3) use of a long-term, sustainable service development perspective, (4) strong service collaboration to avoid wastage, (5) provision of mental health care within primary health care settings, (6) access to services for all, (7) training and supervision and (8) the monitoring of predetermined indicators (i.e., program inputs, processes and outcomes). Consistent with this, WHO is currently developing a global assessment instrument which will measure the capacity of national/regional mental health systems to address mental health and psychosocial needs which emerge following disasters. The WHO-AIMS-E ("Assessment Instrument for Mental Health Systems–Emergency") will allow the identification of the strengths and weaknesses of these systems as a means of furthering their development (WHO Kobe Centre, 2005).

A key consideration of planning and preparation relates to the readiness to enlist community resources for the recovery process. This may include training key community members in health-specific and community-based mental health support processes (Weine et al., 2002). Several authors have highlighted the conceptual distinction between social and mental health interventions following disasters (e.g., Silove, 2005). The primary intention of social interventions is to produce social effects, although they often produce important mental health effects and may therefore support and promote mental health programs. For example, an important early social intervention is the provision of reliable and credible information about the emergency, the plan for response and its progress. This may reduce anxiety in a number of specific ways and enhance the sense of control within a community, particularly during the early stages of recovery. Primary care approaches in several tsunami-affected regions included "befriending" programs, where community members learned active listening and support skills that could be used to assist those experiencing isolation or ongoing distress. Disaster planning should proactively address potential mutual benefits of social and mental health interventions and how they may be implemented to maximize outcomes and avoid competition for limited resources.

A public mental health approach may consider a range of criteria by which to select service priorities, rather than those based upon prevalence alone. It could be said that an implicit strength of effective public health approaches is that they reorientate mental health-care policy towards processes of resilience and natural healing, particularly following disasters, while attempting to maintain resources for vulnerable groups such as those with preexisting mental illness. This issue remains contentious, however de Jong et al. (2003) have proposed core selection criteria for mental health care following disasters which are an attempt to strike such a balance. These include (1) level of functioning, (2) perceived and expressed needs of the population, (3) help-seeking motivations, (4) problems that may be treated with limited resources and (5) cost effectiveness.

Early in 2005, WHO presented 12-month projections for the prevalence of mental health conditions across all the countries affected by the Tsunami (See Table 1). This public health model holds that less severe, higher prevalence conditions, including PTSD, may be responsive to a broader range of psychosocial responses and be more likely to remit without intervention, than low-prevalence, high-disability conditions such as schizophrenia (van Ommeren et al., 2005b). These estimates are noteworthy when compared against the availability of specialist mental health personnel and the number of people requiring mental health care outside the disaster context. In Sri Lanka, for example, WHO estimates

TABLE 1. WHO Mental Health Projections and Recommendations to Populations Affected by the Tsunami. Adapted from: van Ommeren, Saxena & Saraceno (2005b)

Description	Pre-Disaster Prevalence: 12 month Baseline	Post-Disaster Prevalence: 12 month Projected	Aid Response Recommended	Service Sectors Required
Severe Disorder e.g., Psychosis, severe depression, severely disabling form of anxiety disorder	2-3%	3-4%	Mental health care made available in general health services & community mental health services Develop social services to support rehabilitation	Health Services Social Services
Mild to Moderate Mental Disorder e.g., Mild to moderate depression & anxiety disorders, including PTSD	10%	20% Reducing to 15% through natural recovery without intervention	Mental health care made available in general health services & community mental health services Make social interventions and basic psychological support interventions available in the community	Health Services Range of other community services
Moderate to Severe Psychological Distress Does not meet criteria for disorder that may resolve over time or mild distress that does not resolve over time	No estimate	30-50% Reduces over time through natural recovery without intervention	Make social interventions and basic psychological support interventions available in the community	Range of other community services
Mild Psychological Distress Resolves over time	No estimate	20-40% May increase over time as those with severe problems recover	No response	Nil service sector response

that 22,000 to 44,000 tsunami survivors will develop psychological problems serious enough to require longer-term care (cited in Miller, 2005). However, prior to the Tsunami, 384,000 people suffered from severe mental health conditions such as major depression, schizophrenia and bipolar disorder, with up to 2 million afflicted by less severe mental health conditions. Despite such need, the entire country has only 41 psychiatrists (many of whom are academics), eight psychiatric social workers and no psychiatric nurses. Nearly all are situated in the capital, Colombo (WHO, 2001b).

The selection, training and supervision of workers providing mental health assessments and interventions may be key factors determining the integration of mental health into primary health care and the sustainability of established disaster mental health plans (Weine et al., 2002). Such plans need to consider the role of indigenous and traditional health carers. Faith healers, religious and community leaders may provide an important source of culture-specific care and meaning, and generate cohesion within communities. Collaboration between allopathic and traditional service is often advocated but is challenging in practice (Hiegel, 1996). Mental Health practitioners are often very scarce in developing countries, while remote regions affected by disasters may have no such services (WHO, 2001b). In these regions mental health specialists should provide on-the-job training and supervision to general health staff. Community workers may be similarly trained to conduct outreach programs or to provide case-work support to primary care staff. Training such as this is likely to provide the broadest coverage feasible in many regions and allow the identification of severe and potentially long-term mental health disorders. Ongoing, quality supervision provides a vehicle for system-wide capacity building regarding mental health issues as well as the sustainability of many other aspects of the disaster plan (see Weine et al., 2002).

REALITIES: EARLY INTERVENTION AND THE RECOVERY ENVIRONMENT

There remains a shortage of systematic evaluation of the effectiveness of post-disaster mental health interventions. This is due, in part, to the chaos that often ensues following a disaster, but also to the methodological and ethical constraints on research at such times. Anecdotal reports from Tsunami-affected regions included local research teams that deferred research on ethical grounds, to unaligned overseas teams that conducted studies without permission from local authorities or even their nominated institutions. Senior administrators in the region noted the difficult balance between protecting affected groups from intrusive or culturally insensitive inquiry and the "obligation" to conduct systematic studies of impacts and the effectiveness of social and mental health interventions (Miller, 2005). The approaches noted below have had varying levels of scientific review and determinations as to their effectiveness. Some have been endorsed by expert consensus groups (National Institute of Mental Health [NIMH], 2001; Weine et al., 2002; Sphere Project, 2004; National Child Traumatic Stress Network & National Centre for Post Traumatic Stress Disorder [NCTSN & NCPTSD], 2005) due to their observed utility in practice and also for meeting the primary requirement that they "do no harm."

In the immediate aftermath of a disaster, those affected may often appear stunned, bewildered or even dissociated, distressed by events and fearful that more will come. Further risk may result from the desperate search for missing loved ones or the inability to plan or exercise care at such a time. The first priority during and following the emergency is to ensure survival and safety, while also ensuring that the responses themselves do not cause further harm. "Psychological First Aid" is the concept that has been used to describe the type of intervention considered appropriate during this phase and is an important helping response. It is an approach which recognises a post-event "hierarchy of needs" and focuses on the provision of immediate safety and protection from further harm, shelter, support and the triaging of mental health and other needs (Raphael, 1977; NCTSN & NCPTSD, 2005). A key aspect may involve assisting with the whereabouts of loved ones and dealing with knowledge of what may have happened to them. Contact with survivors at recovery centres, for example, may see spontaneous, mutual sharing of their linked experiences and may support their coping. However, this natural sharing should be seen as an informal process and such disclosures should never be actively sought or coerced.

Debriefing approaches and particularly the Critical Incident Stress Debriefing (CISD) approach of Mitchell (1983) were also widely used emergency and early interventions. While its proponents identify it as a technique for emergency workers, it has often been used with disaster-affected populations, including those of the 2004 Tsunami (Miller, 2005). It has been criticised in that it may heighten or maintain arousal levels for some people (Shalev et al., 1998) and that it fails to take account of individual coping patterns and cultural understandings and prescriptions for dealing with such experiences (Litz, Gray, Bryant & Adler, 2002). While it continues to be used, there is no evidence that it is beneficial to either emergency workers or the wider population (Rose, Bisson & Wessely, 2001) or, importantly, that it reduces the risk of PTSD (Litz & Gray, 2004). Some controlled studies have even shown that CISD led to small but significantly worse outcomes (e.g., Bisson,

Jenkins, Alexander & Bannister, 1997). Findings such as these led a recent U.S. Consensus Conference on Mass Violence to recommend against its use with the general population, particularly "single-session" approaches (NIMH, 2001). Conversely, this group has endorsed the wider-population use of Psychological First Aid during the emergency phase and this approach is now internationally recognized (NIMH, 2001; NCTSN & NCPTSD, 2005).

Depending on the type of disaster, a number of survivors may go on to have significant mental health difficulties that may also affect their daily functioning. There is evidence for the effectiveness of a range of mental health interventions in treating conditions that may arise months or even years after a disaster. Depression and anxiety conditions may be common following the tsunami but systems to identify complicated bereavement, alcohol abuse and other presentations, into the longer term, are equally important. Anecdotal reports indicate that many women have experienced profound guilt and anxiety over their inability to save their children when the water struck. Many fishermen have been afraid to return to the sea and attempted to cope through alcohol use. It has been suggested that cultural preferences in some Asian regions may mean that such anguish may be expressed as somatic pain or discomfort and recognised more readily by health workers fluent in local languages and knowledgeable of local culture (Miller, 2005).

In controlled trials, typically in Western countries, Cognitive-Behaviour Therapy (CBT) has been shown to be effective in the treatment of depression, anxiety and specific trauma and likely to benefit those with established post-traumatic conditions following disasters (APA, 2000; Litz et al., 2002). A CBT-based approach to complicated or traumatic grief, "Complicated Grief Therapy" (CGT), has also been shown to be effective in recent controlled trials (Shear, Frank, Houck & Reynolds 2005). However, Tsunami-affected countries may have limited specialist staff available to provide such services, particularly in the earlier stages following the disaster. It is also unclear whether these diagnoses are valid in non-Western countries (van Ommeren, 2003) or that CBT treatments are applicable outside the Western regions in which they were developed (APA, 1994; Oei, 1998). In many countries, the potential direct benefits to these latter groups must also be weighed against the broader provision of training, supervision and planning/coordination of mental health assessment and outreach services. A range of culturally-integrated social interventions have also been shown to have positive mental health effects (Silove, 2005). WHO have emphasised the value, particularly in developing nations, of using limited mental health or other specialist workers to provide broad support to such initiatives, where appropriate, as part of their disaster preparation and response (WHO, 2001a). These workers may also have a key role in designing, training and supervising primary care workers and non-health community workers in "basic" psychological support interventions, such as active listening or problem solving skills (van Ommeren et al., 2005b). This approach has seen significant mental health capacity-building in many countries following the tsunami.

Some early evidence suggests that those affected by the recent Tsunami have shown considerable resilience and are coping better than was expected. It has been suggested that the emphasis in Asian cultures on family and community welfare and support, over individual self-reliance, may have had a powerful positive influence (Miller, 2005). Previous experience of tragedy and upheaval in some regions, including civil conflict in Aceh and Northeast Sri Lanka, may have helped many to develop survival and coping strategies. Many will still be affected in various ways and a range of coordinated social and mental health interventions will need to be implemented. These form part of a wider "recovery environment," the establishment of which would also include disease control, infrastructure and economic resources. The more positive recovery environments actively recognise and support the strengths of individuals and communities to overcome disasters and focus recovery activities on community resources and participation (Sphere Project, 2004).

Social interventions following the Tsunami included the establishment of "child friendly" play spaces and the priority of reopening schools and key businesses. UNICEF was able to support a timely return to school routines in several regions through the distribution of thousands of "school-in-a-box" kits and recreational items. Such programs have the aim of supporting a return to normalcy for children and may also reduce the anxiety and care burden on parents, supporting their own recovery processes and the reestablishment of routines. The reopening of businesses or the establishment of local "micro-economies" (bartering schemes) may be pivotal in stimulating community activity, cohesion and a sense of agency. This may act as a powerful counter to the sense of despair, which can pervade some communities following disaster. While their potential psychological and social benefits may be profound, health and mental health workers may not be familiar with the implementation of social interventions of this kind. This reinforces the need for disaster plans to address the broad range of interventions that may support recovery, including the community workers and networks and the cultural rituals and processes that may be enlisted to achieve these vital outcomes.

CONCLUSION

There is a growing body of science to inform disaster response, both in terms of the epidemiology of mental health conditions and some of the effective interventions that may be provided. This scientific literature is increasingly backed by specific studies in Asian countries which are invaluable in providing understandings appropriate to these cultural settings as well as more broadly. Mental Health service systems may need extra support and guidelines to deliver effective programs for those with both new and existing illnesses that require care as a consequence of disaster. Guidelines such as WHO-AIMS-E and publications such as those of van Ommeren et al. (2005a, 2005b) provide support for both psychological and social interventions, within the wider recovery context. The destruction, injuries and death, as well as ongoing trauma and grief, will all contribute challenges in the aftermath. Nevertheless, primary care models including befriending programs, support for local capacity and ownership of response, may all harness powerful forces of resilience–even in the face of such enormous catastrophes as the Tsunami. The relief of suffering, the compassionate response to those who are injured psychologically as well as physically, opportunities for social rituals and practical support will all help individuals and communities to go forward with courage and ultimately with hope.

REFERENCES

American Psychiatric Association (1994). *Diagnostic and statistical manual of mental disorders*. Washington, DC: American Psychiatric Press.

American Psychiatric Association (2000). Practice guidelines for the treatment of patients with major depressive disorders. *American Journal of Psychiatry*, 157, 1-45.

Birmes, P., Brunet, A., Carreras, D., Ducasse, J. L., Charlet, J. P., Lauque, D. et al. (2003). The predictive power of peritraumatic dissociation and acute stress symptoms for posttraumatic stress symptoms: A three-month prospective study. *American Journal of Psychiatry*, 160(7), 1337-1339.

Bisson, I. J., Jenkins, P. L., Alexander, J., & Bannister, C. (1997). A randomised controlled trial of psychological debriefing for victims of acute harm. *British Journal of Psychiatry*, 171, 78-81.

Bonnano, G. (2004). Loss, trauma, and human resilience. *American Psychologist*, 59, 20-28.

Brewin, C. R., Andrews, B., & Valentine, J. D. (2000). Meta-analysis of risk factors for posttraumatic stress disorder in trauma-exposed adults. *Journal of Consulting and Clinical Psychology*, 68(5), 748-766.

Buckland, J., & Rahman, M. (1999). Community-based disaster management during the 1997 Red River Flood in Canada. *Disasters*, 23(2), 174-191.

de Jong, J. T. V. M., Komproe, I. H., & van Ommeren, M. (2003). Terrorism, human-made and natural disasters as a professional and ethical challenge to psychiatry. *Bulletin of the Board of International Affairs of the Royal College of Psychiatrists*, 1, 8-9.

Dyregrov, A., Kristoffersen, J. I., & Gjestad, R. (1996). Voluntary and professional disaster workers: Similarities and differences in reactions. *Journal of Traumatic Stress*, 9, 541-555.

Engel, G. L. (1961). Is grief a disease? *Psychosomatic Medicine*, 23, 18-22.

Erikson, K. (1994). *A new species of trouble: The human experience of modern disasters*. New York: Norton.

Galea, S., Vlahof, D., Resnick, H., Ahern, J., Susser, E., Gold, J. H., Bucuvalas, M., Gold, J., & Kilpatrick, D. (2003). Trends of probable post-traumatic stress disorder in New York City after the September 11 terrorist attacks. *American Journal of Epidemiology*, 158, 514-524.

Gray, M., Prigerson, H., & Litz, B. (2004). Conceptual and definitional issues in complicated grief. In B. Litz (Ed.), *Early intervention for trauma and traumatic loss* (pp. 65-86). New York: The Guilford Press.

Hiegel, J. P. (1996). Traditional medicine and traditional healers. In J. T. V. M. de Jong & L. Clark (Eds.), *Mental health of refugees*. Geneva: WHO and UNHCR.

Hull, A. M., Alexander, D. A., & Klein, S. (2002). Survivors of the Piper Alpha oil platform disaster: Long term follow-up study. *British Journal of Psychiatry*, 181, 433-438.

Kessler, R. C. (2000). Posttraumatic stress disorder: The burden to the individual and to society. *Journal of Clinical Psychiatry*, 61 (Suppl 5), 4-12.

Kessler, R. C., Sonnega, A., Bromet, E., Hughes, M., & Nelson, C. B. (1995). Posttraumatic stress disorder in the National Comorbidity Survey. *Archives of General Psychiatry*, 52, 1048-1060.

Kokai, M., Fuji, S., Shinfuku, N., & Edwards, G. (2004). Natural disaster and mental health in Asia. *Psychiatry & Clinical Neurosciences*, 58(2), 110-116.

Litz, B., & Gray, M. (2004). Early intervention for trauma in adults: A framework for first aid and secondary prevention. In B. Litz (Ed.), *Early intervention for trauma and traumatic loss* (pp. 87-111). New York: The Guilford Press.

Litz, B., Gray, M., Bryant, R., & Adler, A. (2002). Early interventions for trauma: Current status and future directions. *Clinical Psychology: Science and Practice*, 9, 112-134.

Maes, M., Mylle, J., Delmire, L., & Altamura, C. (2000). Paediatric morbidity and comorbidity following accidental man-made traumatic events: Incidence and risk factors. *Archives of Psychiatry and Clinical Neuroscience*, 250(3), 156-162.

Miller, G. (2005). The Tsunami's psychological aftermath. *Science*, 309(5737), 1030.

Mitchell, J. T. (1983). When disaster strikes. The critical incident stress debriefing process. *Journal of Emergency Services*, 8, 36-39.

Mrazek P., & Haggerty, R. (Eds.) (1994). *Reducing risks for mental disorders: Frontiers for preventative intervention research.* Washington, DC: National Academy Press.

Nakagawa, Y., & Shaw, R. (2004). Social Capital: A missing link to disaster recovery. *International Journal of Mass Emergency and Disasters*, 22(1), 5-34.

Nambi, S. (2005). Conference Presentation: *2nd International Cultural Psychiatry Conference (IAPA).* November 2005, Sydney, Australia.

National Child Traumatic Stress Network & National Centre for PTSD (2005). *Psychological first aid: Field operations guide.* Retrieved December 20, 2005, from http://www.ncptsd.va.gov/pfa/PFA_9_6_05_Final.pdf

National Institute of Mental Health (2001). Mental health and mass violence: Evidence-based early psychological intervention for victims/survivors of mass violence. A Workshop to Reach Consensus on Best Practices. NIH Publication No. 02-5138. Washington, DC: US Government Printing Office.

Norris, F. (2005). Range, magnitude and duration of the effects of disaster on mental health: Review update 2005. Dartmouth College Research Education Review Article. Retrieved December 12, 2005, from http://www.redmh.org/research/general/REDMH_effects.pdf

Norris, F., Friedman, M., Watson, P., Byrne, C., Diaz, E., & Kaniasty, K. (2002). 60,000 disaster victims speak, Part I: An empirical review of the empirical literature 1981–2001. *Psychiatry*, 65, 207-239.

North, C. S., Nixon, S., Hariat, S., & Mallonee, A. (1999). Psychiatric disorders among survivors of the Oklahoma City bombing. *Journal of the American Medical Association*, 282, 755-762.

Oei, T. (1998). *Behaviour therapy and cognitive behaviour therapy in Asia.* Sydney: Edumedia.

Phifer, J., & Norris, F. (1989). Psychological symptoms in older adults following disaster: Nature, timing, duration and course. *Journal of Gerontology*, 44, 201-217.

Pynoos, R., & Nader, K. (1988). Psychological first aid and treatment approach to children exposed to community violence: Research implications. *Journal of Traumatic Stress*, 1, 445-473.

Quarantelli, E. L. (Ed.) (1998). *What is disaster? Perspectives on the question.* London: Routledge.

Raphael, B. (1977). The Granville train disaster: Psychological needs and their management. *Medical Journal Australia*, 1(9), 303-305.

Raphael, B., Dunsmore, J., & Wooding, S. (2004). Terror and trauma in Bali: Australia's mental health response. *Journal of Aggression, Maltreatment and Trauma,* 9(2), 245-256.

Raphael, B., Martinek, N., & Wooding, S. (2004). Assessing loss, psychological trauma and traumatic bereavement. In J. P. Wilson & T. M. Keane (Eds.), *Assessing psychological trauma and PTSD–Second Edition* (pp. 492-512). New York: The Guilford Press.

Raphael, B., & Wilson, J. P. (1993). Theoretical and intervention considerations in working with victims of disaster. In J. P. Wilson & B. Raphael (Eds.), *International handbook of traumatic stress syndromes* (pp. 105-117) . NewYork: Plenum Press.

Raphael, B., & Wooding, S. (2004). Early mental health interventions for traumatic loss in adults. In B. Litz (Ed.), *Early intervention for trauma and traumatic loss* (pp. 147-178). New York: The Guilford Press.

Regehr, C., & Sussman, T. (2004). Intersections between grief and trauma: Toward an empirically based model for treating traumatic grief. *Brief Treatmentand Crisis Intervention,* 4, 289-309.

Rose, S., Bisson, J., & Wessely, S. (2001). Psychological debriefing for preventing post traumatic stress disorder (PTSD) (Cochrane Review). *The Cochrane Library,* 4, Oxford, U.K.: Update Software Ltd.

Salama, P., Spiegel, P., Talley, L., & Waldman, R. (2004). Lessons learned from complex emergencies over the past decade. *Lance,* 364(9447), 1801-1813.

Salter, J. (1995). Towards a better disaster management methodology. *Australian Journal of Emergency Management,* 10(4), 8-16.

Sattler, D.N., Preston, A. J., Kaiser, C. F., Olivera, V. E., Valdez, J., & Schlueter, S. (2002). Hurricane Georges: A cross-national study examining preparedness, resource loss, and psychological distress in the U.S. Virgin Islands, Puerto Rico, Dominican Republic, and the United States. *Journal of Traumatic Stress,* 15, 339-350.

Shalev, A. Y., Sahar, T., Freedman, S., Peri, T., Glick, N., Brandes, D., Orr, S. P., & Pitman, R. K. (1998). A prospective study of heart rate responses following trauma and the subsequent development of posttraumatic stress disorder. *Archives of General Psychiatry,* 55, 553-559.

Shear, K., Frank, E., Houck, P. R., & Reynolds, C. F. (2005). Treatment of complicated grief: A randomized controlled trial. *Journal of the American Medical* Association, 293, 2601-2608.

Shore, J., & Tatum, L. (1986). Psychiatric reactions to disaster: The Mount St. Helens experience. *American Journal of Psychiatry,* 143, 590-595.

Silove, D. (1999). The psychosocial effects of torture, mass human rights violations, and refugee trauma: Toward an integrated conceptual framework. *Journal of Nervous & Mental Disease,* 187(4), 200-207.

Silove, D. (2005). The best immediate therapy for acute stress is social. *Bulletin of World Health Organization,* 83, 75-76.

Silverman, G. K., Johnson, J. G., & Prigerson, H. G. (2001). Preliminary explorations of the effects of prior trauma and loss on risk for psychiatric disorders in recently widowed people. *Israel Journal of Psychiatry and Related Sciences,* 38, 202-215.

Sphere Project (2004). Humanitarian charter and minimum standards in disaster response. Geneva: Sphere Project. Retrieved December 22, 2005, from http://www.sphereproject.org/handbook/index.htmj

Spiegel, P. (2005). Differences in world responses to natural disasters and complex emergencies. *Journal of the American Medical Association,* 293(15), 1915-1918.

Summerfield, D. (2005). What exactly is emergency or disaster "mental health"? *Bulletin of World Health Organization,* 83, 76.

Tedeschi, R. G., & Calhoun, L. G. (2004). Post-traumatic growth: Conceptual foundations and empirical evidence. *Psychological Inquiry,* 15, 1-18.

Udomratm, P. (2005). *Conference Presentation: 2nd International Cultural Psychiatry Conference (IAPA).* November 2005, Sydney, Australia.

Ursano, R. J., Fullerton, C. S., & Norwood, A. E. (2003). *Terrorism and disaster: Individual and community mental health interventions*. Cambridge: Cambridge University Press.

van Ommeren, M. (2003). Validity issues in transcultural epidemiology. *The British Journal of Psychiatry,* 182, 376-378.

van Ommeren, M., Saxena, S., & Saraceno, B. (2005a). Mental and social health during and after acute emergencies: Emerging consensus? *Bulletin of World Health Organization*, 83, 71-75.

van Ommeren, M., Saxena, S., & Saraceno, B. (2005b). Aid after disasters: Needs a long term public mental health perspective. *British Medical Journal*, 330, 1160-1161.

Weine, S., Danieli, Y., Silove, D., van Ommeren, M., Fairbank, J. A., & Saul, J. (2002). Guidelines for international training in mental health and psychosocial interventions for trauma exposed populations in clinical and community settings. *Psychiatry*, 65(2), 156-164.

WHO Kobe Centre: World Health Organization Centre for Health Development (2005). Consultative Meeting on the Development of a WHO Assessment Instrument for Mental Health Systems–Emergencies (WHO-AIMS-E). Kobe: WHO. Retrieved December 22, 2005, from http://www.who.or.jp/2005/mentalhealth.html

World Health Organisation (2001a). Mental health: New understanding, new hope. World Health Report 2001. Geneva: WHO.

World Health Organisation (2001b). Atlas of mental health resources in the world. Geneva: WHO.

World Health Organisation (2003). Mental health in emergencies: Mental and social aspects of health of populations exposed to extreme stressors. Department of Mental Health and Substance Dependence. Geneva: WHO.

doi:10.1300/J198v05n03_01

Disability and Disaster Recovery: A Tale of Two Cities?

Mark Priestley
Laura Hemingway

SUMMARY. This paper examines the connections between disability and disaster from a global perspective. Concepts from the research and policy literature are used to distinguish between individual and social models of disability, and between natural hazards and human disasters. These concepts are then employed to investigate data on the response to disabled people's recovery needs in two recent case studies: the Asian tsunami and Hurricane Katrina. The analysis combines primary, secondary and tertiary sources to explore disability issues in the reconstruction of inclusive communities and the lessons that may be learned about disaster preparedness in poor communities. The conclusions suggest that more attention should be paid to social model approaches, particularly in understanding global links with poverty, and that disabled people's organisations should be resourced as agents of disaster recovery and preparedness. doi:10.1300/J198v05n03_02

Mark Priestley, PhD, is Reader in Disability Studies, Centre for Disability Studies, University of Leeds, LS2 9JT UK (E-mail: m.a.priestley@leeds.ac.uk).

Laura Hemingway, MA, is a Doctoral Research Student, Centre for Disability Studies, University of Leeds, LS2 9JT UK.

[Haworth co-indexing entry note]: "Disability and Disaster Recovery: A Tale of Two Cities?" Priestley, Mark, and Laura Hemingway. Co-published simultaneously in *Journal of Social Work in Disability & Rehabilitation* (The Haworth Press, Inc.) Vol. 5, No. 3/4, 2006, pp. 23-42; and: *Asian Tsunami and Social Work Practice: Recovery and Rebuilding* (ed: Ngoh Tiong Tan, Allison Rowlands, and Francis K. O. Yuen) The Haworth Press, Inc., 2006, pp. 23-42. Single or multiple copies of this article are available for a fee from The Haworth Document Delivery Service [1-800-HAWORTH, 9:00 a.m. - 5:00 p.m. (EST). E-mail address: docdelivery@haworthpress.com].

Available online at http://jswdr.haworthpress.com

doi:10.1300/J198v05n03_02

KEYWORDS. Disability, disaster planning, tsunami, hurricane Katrina

As the catastrophic events of 2004-2005 so graphically illustrated, natural hazards affect millions, causing devastation to lives, livelihoods and communities. In counting the human cost the disability dimension has been something of a hidden story. Disability issues have been both underreported and of limited scope within research on disaster recovery, with a tendency towards individualised psychological interventions at the expense of social model approaches. As we argue in this paper, the success of long-term recovery for disabled people depends on redressing this imbalance.

During the past 30 years there has been a paradigm shift in the global discourse and politics of disability, initiated through the activism of the international disabled people's movement (Campbell & Oliver, 1996; Charlton, 1998; Driedger, 1989; Fleischer & Zames, 2001). The traditional view within social science and rehabilitation was to view the disadvantage associated with disability as an individual problem caused by impairment. From this perspective, the most appropriate response was either to correct the impairment or to help the person "come to terms" with it, by negotiating different (less valued) social roles. By contrast, social interpretations of disability have shown how people with similar impairment characteristics become more or less "disabled" in different environments and social circumstances (Ingstad & Reynolds Whyte, 1995; Zola, 1989). This allows us to view disability as a social problem resulting from social processes (Priestley, 2005).

The shift of emphasis, from the individual to the social, has allowed activists, academics, practitioners and policymakers to engage in a far-reaching critique of the ways in which people with perceived impairments are affected by different social arrangements, and to envisage more inclusive alternatives. Thus, in his ground-breaking book *Social Work with Disabled People*, Oliver (1983) distinguished between interventions based on "individual" and "social" models of disability. Extending these concepts, Oliver (1990, 1996) defines individual model approaches as underpinned by a personal tragedy theory of disability, predominantly characterised by psychological and medicalised interventions that seek remedy in the person. By contrast, social model interventions seek remedy in challenges to social policies, practices and institutions. There are now few debates within the contemporary disability studies literature that are not framed by this conceptual distinction (see Albrecht, Seelman, & Bury, 2001; Barnes & Mercer, 2005; Swain, French, Barnes, & Thomas, 2004).

NATURAL HAZARD OR HUMAN DISASTER?

Natural hazards are often classified into three main types (e.g., Frampton, McNaught, Hardwick, & Chaffey, 2000). These may be characterised as geological/tectonic (for example, earthquakes or volcanoes, which may create tsunamis), climatic (which may lead to floods, droughts and hurricanes) and geomorphological/land instability hazards (for example, landslides or avalanches). However, it is important to distinguish between these natural "hazards" and human "disasters." Natural hazards lead to disasters when they damage human lives and environments. Although natural hazards may not be eradicated, human disasters may be lessened by reducing vulnerabilities, particularly for poor or socially excluded groups.

In the context of this paper, our underlying hypothesis is that natural hazards are realised disproportionately as human disasters for disabled people, and particularly for disabled people in low-income communities. Since disabled people are consistently amongst the poorest of the poor in both high-income and low-income countries, the realisation that human disasters are socially created has a particular resonance for thinking about disability and disaster recovery. Just as disability is not the inevitable consequence of physical or cognitive impairment, disaster is not the inevitable consequence of natural hazard.

Synthesising concepts from disability and disaster studies requires an approach that treats disability as more than an individual or therapeutic concern, and that considers the impact of disabling environments and relationships on disaster survivors. Since the social model approach owes its vigour to the global activism of disabled people, it is also important to consider the role of activism in developing inclusive policies and practices. Given the principle of "nothing about us without us" (Charlton, 1998), the significance of disabled people and their organisations as agents of disaster recovery cannot therefore be overlooked.

Turning to the disaster literature there is surprisingly little evidence that such debates have so far impacted on social work and rehabilitation research. While critical perspectives exist in policy and practice documents, discussion in the academic literature is extremely limited. For example, a systematic review of journal abstract databases revealed 180 references to disability and disaster (i.e., by searching for terms such as disaster, natural hazard, tsunami, earthquake, hurricane or terrorist in combination with variations on the terms impairment, disability and handicap). After eliminating articles of no direct relevance, four broad themes are evident. By far the most frequently addressed issue is that of measuring and treating psycho-emotional impairments arising from disaster events, in particular post-traumatic stress disorder (e.g., Chang, Connor, Lai, Lee, & Davidson, 2005; Livanou et al., 2005). Similarly, there are epidemiological and medical studies examining the incidence and treatment of acquired physical impairment (e.g., Matsuoka et al., 2002; Naghi et al., 2005). Third, but to a lesser extent, there are evaluations of the vulnerability of disabled people to natural hazards or terrorist attacks (e.g., Chou et al., 2004; Osaki & Minowa, 2001). Fourth, there are a small number of studies addressing the social impact of disasters on disabled people (e.g., Duyan & Karatas, 2005; Gignac, Cott, & Badley, 2003; Takahashi, Watanabe, Oshima, Shimada, & Ozawa, 1997).

The bulk of this literature addresses disability from an individual model perspective. Moreover, interventions are characterised overwhelmingly as the province of medical or other therapeutic professionals and there is little reference to the contribution of disabled people in recovery planning or practice. Looking beyond the academic journals, there is more evidence of alternative approaches that address social model concepts or that consider the value and agency of disabled people's organisations as partners in planning and recovery (e.g., Blanck, 1995; Center for International Rehabilitation, 2005; Kett, Stubbs, & Yeo, 2005; Wisner, 2002). The remainder of this paper examines the extent to which such themes are evident in recovery from two recent and globally significant disaster events.

CASE STUDIES AND METHODS

The tsunamis that devastated coastal areas across South Asia resulted from the Sumatra-Andaman Islands earthquake beneath the Indian Ocean on December 26, 2004. Although this was only the fourth largest earthquake since 1900, the magnitude of the human disaster was greater than any similar event in recorded history. According to current estimates around 275,000 people were killed in 11 countries, the majority in Indonesia. One year after, more than a million people remained displaced (US Geological Survey, 2005). In terms of the definitions outlined earlier, this unprecedented human disaster arose from the coincidence of a substantial tectonic hazard and the extreme vulnerability of certain coastal communities.

The impact of Hurricane Katrina appears somewhat more limited yet its significance should not be underestimated. Although Katrina was not the most powerful Atlantic Basin storm of 2005 (both Wilma and Rita were stronger) it resulted in more widespread and costly human disaster than any single natural hazard in United States history. Between August 25 and 29 the storm impacted on Florida and the Gulf States, affecting an area the size of Great Britain. Extensive damage was caused by the accompanying storm surge, notably in the flooding of the city of New Orleans. According to official estimates around 1,300 people were killed, more than a million were displaced from their neighbourhoods and up to five million lost electrical power.

In order to investigate recovery for disabled people in such recent disasters it is necessary to draw on a range of data sources. First, there is a rapidly growing, but largely unmapped, "grey literature" of research and evaluation produced by practice or advocacy-based organisations. Considerable insight was also gained from textual analysis of policy and guidance documents produced by governmental and non-governmental agencies (e.g., in published reports, emergency guidance, organisational websites, etc.).

Much was learned from the omission of disability as well as from its inclusion and it was important to systematically review documents from government agencies, donors, relief agencies and disability organisations. These were collected from personal contacts, reviews and detailed Internet searches. It is important to note that these reviews were conducted between September and December 2005 and that new documents were emerging throughout this period. Media reports, press releases and online discussion forums were also helpful in identifying how disabled people were affected. In addition, primary data was generated through e-mail or telephone contact with key informants in 18 organisations involved in the tsunami or Katrina recovery effort. These included: international coordinating organisations, international donor organisations, disabled people's organisations, US government and state agencies, and local service providers or disability projects.

In an era of economic and cultural globalisation, the realities of disasters are rapidly shared via media coverage and Internet communication technologies. In the aftermath of the Asian tsunami and Hurricane Katrina, patterns of global connectivity also affected patterns of response at disaster sites. Data from this global information flow was important because connectedness creates opportunities for (a) knowledge sharing and the targeting of aid or expertise; (b) private transfers of information and resources that are not evidenced in the mainstream literature; (c) peer-to-peer organisational relief and support beyond traditional donor networks. This kind of connectedness has specific consequences for disabled people, as survivors and as agents of recovery. In particular, it illustrated how the international disabled people's movement is challenging traditional models of disaster relief and traditional relationships of power between "professionals" and communities.

While disability equality issues have been historically marginalised, they are increasingly referenced in disaster evaluation and practice development. Major relief agencies and international NGOs have comm- issioned disability audits in their post-tsunami evaluations and increased attention is being paid to the voice of disabled people. It is not our intention to replicate these studies but to examine the available data in its theoretical and global policy context. Our analysis here focuses primarily on recovery in the "reconstruction" or "rehabilitation" phase, and on preparedness for future events. Elsewhere, we explore in detail the initial impact on disabled people of the Tsunami and Katrina, and the barriers they experienced in accessing immediate relief, evacuation and shelter (see Hemingway & Priestley, 2006, forthcoming).

REBUILDING INCLUSIVE COMMUNITIES

Interventions for long-term recovery are inevitably diverse. They may be restorative or preventative, and involve a range of actors in one-off operations, complex long-term programmes or projects targeted at particular groups (e.g., Perrin, 1998; Smith, 2001). In the case study examples massive investment has been committed to rebuilding lives, communities and infrastructure. If disabled people's needs are to be met it is imperative that reconstruction integrates accessibility and that recovered communities are inclusive communities. Considering this "rehabilitation" from a social model perspective involves looking beyond individualised and therapeutic interventions to consider the barriers to full participation and equality.

Breakdown of Infrastructure

Breakdown of infrastructure affects the whole community but there are additional barriers for disabled people in accessing communication systems, the built environment, transport, employment, education, health care, public services and so on (California Department of Rehabilitation, 1997; Shaw & Goda, 2004; World Health Organisation, 2005). Disabled people may be disproportionately disadvantaged by depletion of both physical infrastructure and human services. Thus:

> The aftermath of Hurricane Katrina has led to large-scale displacement that has interrupted the networks of support that individuals with disabilities have. People will need knowledgeable help in arranging essential services in new environments with limited contacts and little knowledge of local resources. (National Organization on Disability [NOD], 2005, p. 12)

There was considerable concern amongst key informants in Louisiana about the deterioration of life choices for disabled people who had previously lived independently, with established networks of community support, but who now found themselves in residential nursing homes. The issue here is not simply damage to the physical environment but the breakdown of social networks and social capital (see Nakagawa & Shaw, 2004). The implications are somewhat different in richer and poorer communities and in countries with different types of social infrastructure. Where there are highly developed public support services for disabled people, breakdown of infrastructure can be sudden and damaging. In the short term there may be significant loss of independence and a return to more medicalised forms of care but there is also the capacity for relatively rapid restoration or relocation of familiar independent living solutions.

Where networks of support are reliant upon informal unpaid care things may be better or worse. On the one hand, there can be great resilience if informal networks survive the disaster, buffering the breakdown of formal structures. On the other hand, the consequences can be catastrophic, and potentially fatal, where disabled people are separated from supportive relationships. In many tsunami-affected areas there has been great concern for disabled people separated or bereaved from family members who previously provided financial and/or physical assistance. Commenting on the challenges in India's reconstruction:

> The government should also be sensitive to the fact that disabled people can be made destitute not only when they lose their own livelihoods but when relatives who in the past provided care or assistance are killed or can no longer afford to help them. (Human Rights Watch, 2005, p. 78)

Scarcity of resources after a disaster can intensify discriminatory attitudes and behaviour within families and communities, leading to further exclusion. Disabled children, including those with newly acquired impairments, may be excluded from education. The Interagency Network for Education in Emergencies argues that recovery interventions need to address negative cultural perceptions of disability, to "ensure that schools are prepared, and that facilities are accessible" and that teachers are "well versed in principles of social justice, diversity and inclusion" (Interagency Network for Education in Emergencies [INEE], 2005a). Amongst the concerns they raise:

> Children can be expelled from their former schools and their right to education is addressed only in the final phases of the "reconstruction" process. Moreover, if they eventually are to have access to education, it is often in institutions and special schools far away from their former classmates . . . (INEE, 2005b)

Significantly, where disabled people have been poorly supported in the past the impact is likely to be increased. For example, in the Andaman and Nicobar Islands disabled survivors were doubly disadvantaged in recovery due to the low levels of supportive infrastructure existing prior to the tsunami event (Handicap International, 2005a). Thus, initial assessments highlighted the absence of local rehabilitation services, disability NGOs, or inclusive educational opportunities (Chari & Padmanabhan, 2005). Similarly, Western rehabilitation teams working in Aceh noted with apparent surprise the lack of facilities for producing affordable, modern artificial limbs in Indonesia.

The tsunami recovery effort has brought NGOs into contact with the stark reality of life and poverty for disabled people. For example, initial interventions by Handicap International (2005b) focused on nursing, trauma and physiotherapy. As a consequence, the extent of preexisting mortality risks for disabled people in Sri Lanka, and the lack of therapeutic services, became apparent. This realisation led to establishing new capacity building projects in affected areas, such as CBR training for social workers and participation in "access for all" initiatives with local organisations of disabled people (Handicap International, 2005c).

As these examples begin to show, it is essential to view disaster recovery for disabled people from a social model perspective. There have been great and urgent needs for medical and therapeutic interventions, and much effort has been expended in addressing physical and psychological trauma. Yet such interventions have quickly revealed underlying structural weaknesses in the support and integration available to disabled people in poor communities.

The Challenge of Inclusive Reconstruction

Devastation of infrastructure creates opportunities as well as barriers, providing an unexpected catalyst for social change (Hastie, 1997; Keen, 2001; Oosters, 2005). Reconstruction offers opportunities to improve the lives of disabled people, to assert social rights more clearly, and to facilitate social and economic integration. Indeed, some commentators have pointed to the emancipatory potential of the recovery process, provided investments can be targeted towards inclusion. For example, in relation to reconstruction in Louisiana:

> The immediate and long-term rebuilding process offers a unique opportunity to build, on an unprecedented scale, accessible communities and accessible and adaptable housing. This will help thousands of people with disabilities maintain or improve their ability to live independently and will enable hundreds of thousands of people, regardless of disability, to age-in-place as they acquire activity limitations. (NOD, 2005, p. 14)

Funding long-term recovery is a mammoth task and investment has come from many sources. Unsurprisingly, the degree to which such funding is contingent upon the inclusion of disabled people varies considerably. Organisational approaches to disability equality often reflect the protection afforded in the donor country. For example, the allocation of more than US$650 million to USAID's Tsunami Recovery and Reconstruction Fund comes with certain conditions. Drawing on the principles of the 1989 Americans With Disability Act and the 1968 Architectural Barriers Act, USAID's disability policy promotes the early engagement of disabled people's organisations in recovery and seeks:

> to avoid discrimination against people with disabilities in programs which USAID funds and to stimulate an engagement of host country counterparts, governments, implementing organizations and other donors in promoting a climate of nondiscrimination against and equal opportunity for people with disabilities. (USAID, 1997, p. 2)

USAID was among a number of organisations to establish donor-co-operation on disability issues under the auspices of the World Bank and maintains its own disability team, together with a specific policy on standards of accessibility for funded construction projects (USAID, 2004). In principle, this means that its contractors must comply with American disability accessibility standards (except where discretionary waivers are granted). By comparison, the UK Disasters Emergency Committee does not explicitly require relief agency members (currently 13 NGOs) to comply with accessibility standards. However, organisations are required to be "committed" to the Sphere charter and minimum standards, which address disability equality in the context of nondiscrimination more generally (Sphere Project, 2004).

In Louisiana, federal, state and local building codes are impacting on environmental planning and reconstruction, whereas in many tsunami affected areas there has been little systematic engagement with access standards. However, national differences are apparent with stronger calls for accessibility in countries where there is existing disability legislation, such as the Disability Act in India, compared to countries, such as Indonesia, with less specific legal provisions (e.g., Chari & Padmanabhan, 2005). Considering the diversity of national non-discrimination laws (Quinn & Degener, 2002), regulation alone cannot ensure the inclusion of disabled people in recovering communities. The unprecedented level of funding for reconstruction in South Asia brings the possibility of far more accessible communities, yet there is a fear that local populations will not benefit to the same degree as wealthy tourists from the West. Such inequalities will be particularly acute in poor coastal communities where, as Yeo (2005) notes, different expectations may apply to accessibility in the construction of tourist hotels and local housing stock.

Structural inequality and poverty in disaster zones will never be overcome through the standardisation of non-discrimination policies alone (although current moves towards a UN Convention affirms disabled people's global human rights). In order to ensure that revived communities are also inclusive communities it is essential to bring expertise on disability discrimination, disabling barriers and independent living solutions directly into the recovery process. Communities with higher levels of social capital and participation make more successful and rapid recoveries (e.g., Nakagawa & Shaw, 2004) yet the social capital that exists within the disabled people's movement is not well-integrated. It is therefore essential that disabled people's organisations are included in the "social domains" where disaster experts, governments and local people exchange ideas and practices (Hilhorst, 2003). If the goal is inclusion then the process of recovery must reflect this principle, too. As Betts (2005) argues, disabled people's organisations must therefore be considered as:

> . . . best placed to provide advice on the needs of people with disabilities, as well as to help make sure they are part of the rebuilding and long-term future of their communities.

In both case studies, disability organisations found it difficult to engage with emergency management teams and experienced rapid depletion of resources, undermining capacity for longer-term involvement. In addition, early evaluations suggest that mainstream interventions were based primarily within an individual (medical) model approach and that there was a failure to incorporate social model or rights-based alternatives (e.g., NOD 2005). As Kett et al. (2005, pp. 8-9) also conclude:

> It was striking that more agencies now use the language of social model and inclusion, but have misunderstandings and do not really apply it in practice. Disabled persons still tended to be lumped together under the heading of "vulnerable groups" rather than being perceived as rights-holders.

The data on immediate relief capacity suggests that social model and rights-based interventions were more likely to be found in the peer-to-peer responses of disabled people's organisations and their immediate allies (Hemingway & Priestley, 2006, forthcoming). This finding underlines the significance of resourcing and involving these organisations for disaster preparedness. The available evidence suggests that local disabled people and their organisations have been consulted in some recovery projects, but that this engagement has been neither widespread nor prioritised. At the international level, some organisations, such as World Vision and the World Bank, have targeted consultation with disabled people while others have not. Disabled people's organisations possess considerable untapped social capital that is vital to the construction of inclusive communities, and interventions could be made more effective by prioritising their direct and sustained engagement.

PREPARING FOR THE FUTURE

To conclude, we examine the extent to which social model perspectives and the engagement of disabled people are evident in disaster planning. Ultimately, disaster preparedness must go beyond practical guidance if it is to address the underlying structural inequalities that disadvantage disabled people in disaster-prone communities, particularly in poor communities. Broadly speaking, existing guidance can be considered under two headings: that which deals specifically with disability issues and that which is generic but may refer to disability. International aid agencies and human rights groups are beginning to recognise the extent of disabled people's vulnerability to natural hazards, and their exclusion from relief and recovery, but these concerns have yet to be mainstreamed in disaster preparedness.

The International Decade for Natural Disaster Reduction was established in 1989 (UN General Assembly resolution 44/236) without reference to particular population groups. Subsequently, the *Yokohama strategy and plan of action for a safer world* noted a lack of success in protecting "poor and socially disadvantaged groups," particularly in developing countries (International Strategy for Disaster Reduction, 1994). Taking into account the structural link between poverty and vulnerability, an important principle of the strategy was to promote disaster preparedness as integral to global development policy. In order to achieve this, increased awareness amongst vulnerable communities, active local participation in risk management, and a greater involvement of indigenous NGOs would be required, in order to:

> stimulate genuine community involvement and empowerment of women and other socially disadvantaged groups at all stages of disaster management programmes in order to facilitate capacity building, which is an essential precondition for reducing vulnerability of communities to natural disasters. (International Strategy for Disaster Reduction, 1994, p. II.17)

More than a decade later, as the tsunami disaster was unfolding in January 2005, the World Conference on Disaster Reduction adopted the *Hyogo Framework for Action 2005-2015*. There is as yet only passing reference to disabled people in this initiative and no recommendation on how proposed "social safety-net mechanisms" might be achieved (United Nations, 2005). Concerns raised in our research, particularly from key informants involved in the relief effort, suggest that this is unsurprising and that there remains a distinct lack of general awareness about disability issues in disaster preparedness.

In principle, disabled people should receive equal access to support. For example, UNCHR's Guiding Principles on Internal Displacement apply without discrimination, including disability discrimination, and acknowledge the need to take full account of "special needs" without delay or distinction (United Nations Commission on Human Rights, 1998). The Sphere Project handbook includes minimum standards for non-discrimination and specific reference to disabled people in each chapter–in relation to community involvement/consultation, nutritional needs, access to water and sanitation, vulnerability to sexual exploitation, provision of clothing, bedding, utensils, and access to health care (Sphere Project, 2004).

However, in the majority of planning documents disabled people are identified only as members of "vulnerable" or "special needs" populations, while specific references are few and limited. There is little acknowledgement of disabled people's heterogeneity. While some make passing reference to the participation of disabled people (e.g., World Health Organisation [WHO], 1999) there are significant publications that fail to mention disability at all. For example, the Code of Conduct for The International Red Cross and Red Crescent Movement and NGOs in Disaster Relief contains no reference to discrimination on grounds of disability, referring only to "race, creed or nationality" and to women (International Federation of Red Cross and Red Crescent Societies [IFRCRCS], 1994).

Where barriers for disabled people are highlighted these focus on a narrow range of activity limitations (e.g., in access to sanitation facilities). However, several documents now direct the reader to additional advice on managing relief for specific groups, such as the FEMA publication *Assisting People with Disabilities in a Disaster*. Whilst such advice is practical and useful, it rarely challenges the assumption that disabled people are vulnerable and dependent. In this sense, there needs to be a move away from perceptions that disabled people are simply helpless victims and towards an understanding that they may be "capable of assuming responsibility" (Sustainable Environment and Ecological Development Society, 2001, p. 58).

In recent years, a new wave of preparedness literature has provided more specific resources for disabled people and relief agencies. A significant difference in these sources is that there is a greater recognition of disabled people's resilience and agency in mobilising personal resources or social networks. However, the increasing focus on personal responsibility for risk management does not adequately address the need for structural interventions. There should therefore be some concern about encouragement to devolve risk responsibility to disabled people and their families unless there is a corresponding engagement with building resilient, inclusive and disaster sustainable communities for the future.

The case study data suggests some surprising similarities between the Asian tsunami and Hurricane Katrina (although there are important qualitative distinctions in terms of scale and impact). In both cases, recovery evaluations have identified shortcomings by governments, disaster planners, donor organisations, relief agencies, shelter managers, and construction contractors in responding adequately to the needs of disabled survivors. Given the unprecedented scale of the tsunami impact, and the structural challenges of implementing disability equality in low-income countries with high exposure to natural hazards, it may be less surprising that disability issues have slipped on the recovery agenda in South Asia. It is perhaps more surprising that continued criticism has been levelled at disaster response in the USA, and this raises some difficult questions.

A significant factor in the level of disaster preparedness in the USA arises from the historic vibrancy of political activism amongst disabled people's organisations in California (dating from the early 1970s). Rising political sensitivity to equality issues combined with the natural hazards posed by the San Andreas Fault appear to have led local disability activists to a much earlier engagement with disaster preparedness than elsewhere in the world. For example, Tierney, Petak, and Hahn's (1988) book, *Disabled Persons and Earthquake Hazards,* demonstrated early potential for a systematic and multi-model analysis of these issues. Similarly, the 1994 Northridge Earthquake led to the establishment of Disabled People and Disaster Planning (DP2, 2001) with recommendations on accessibility planning and guidance for shelter managers, rescue workers and services.

Reactions to the 9-11 attacks on the World Trade Centre provided a further catalyst for preparedness activity, in the establishment of the National Organization on Disability's Emergency Preparedness Initiative and the National Center on Emergency Planning for People with Disabilities. More recently, in July 2004, the US Government established an "Interagency Coordinating Council on Emergency Preparedness and Individuals with Disabilities" under the auspices of the Department of Homeland Security. Given this high level of commitment and resource it is tempting to assume that disabled people in the

USA would have been exceptionally well protected and prepared for the impact of Hurricane Katrina. The reported "systematic failure" of disaster planners to address disabled people's needs and the specific difficulties encountered by individuals and organisations in Louisiana need therefore to be explained if they are to be avoided in future (NOD 2004, 2005).

CONCLUSIONS

The arguments presented in this paper were derived from a systematic review of the disability and disaster literature, and an examination of diverse data arising from two recent case studies. We began by noting similarities between concepts of human disaster and the social model of disability, distinguishing between "natural" occurrences and socially produced vulnerabilities (i.e., between "natural hazards" and "human disasters" or between individual "impairments" and social inequalities characterised as "disability"). At the intersection of these separate debates there is a useful synthesis that helps to explain the specific vulnerability of disabled people in disaster scenarios. Indeed, the historic and structural forces of global production, consumption, power and poverty that create vulnerability in poor communities throughout the world are essentially the same forces that exacerbate the vulnerability of disabled people. Where there is exclusion from full participation and equality there is a human disaster waiting to happen, and while disabled people remain amongst the poorest of the poor they will also be amongst the worst affected.

These differential impact patterns can be seen in disaster evaluation reports and in the illustrative case study examples. Disabled people are less likely to be included in reconstruction or to be considered within the mainstream of disaster planning. These risks are increasingly acknowledged within the disaster community but reality does not yet match the rhetoric. Local evaluations reveal specific implementation gaps (such as a lack of commitment, lack of knowledge, lack of guidance, lack of communication or lack of resources) and it is clear that more could be done to communicate directly with disabled people and their organisations.

There is huge untapped potential and readiness within such organisations and they should be considered as key partners in disaster preparedness (although they will require investment and internal capacity building to fulfill this role). These practical issues could be easily addressed in disaster planning, relief and recovery, and there is an increasing range of available guidance from dedicated disability projects. Linking this more effectively into generic disaster preparedness would assist in ensuring that disabled people are not overlooked. However, recovery and planning initiatives must move beyond individual and therapeutic interventions to consider disability in terms of human rights and structural exclusion.

When viewed in a global perspective the evidence suggests that, even where established systems and knowledge exist, disabled people face a greater risk of death, injury, displacement, destitution or loss of autonomy. Whilst there are stark differences in the response capacity of low-income and high-income countries (e.g., between Indonesia and the USA) there are also unsettling similarities in the experiences of disabled people from poor, predominantly black, communities (whether in Aceh or Louisiana). In this sense, the boundaries of the excluded "majority world" continue to be less defined by geography than by poverty and structural inequality. Understanding the connection between disability and disaster involves understanding the connection between disability, poverty and social exclusion. Ensuring that recovered communities are also inclusive communities requires a greater understanding of the social model and a greater engagement with disabled people's organisations, but above all recovery interventions that facilitate the full participation and equality of disabled people in all aspects of social development and poverty reduction.

REFERENCES

Albrecht, G. L., Seelman, K. D., & Bury, M. (Eds.). (2001). *The handbook of disability studies*. London: Sage.

Barnes, C., & Mercer, G. (Eds.) (2005). *The social model of disability: Europe and the Majority World*. Leeds: Disability Press.

Betts, J. (2005). *Disabled people and the Tsunami*. World Vision UK. Retrieved November 2005, from the World Wide Web: http://www.worldvision.org.uk/tsunami/viewpoints/disability.asp

Blanck, P. (1995). *Disaster mitigation for persons with disabilities: Fostering a new dialogue*. Washington, DC: Annenberg Washington Program in Communications Policy Studies of Northwestern University.

California Department of Rehabilitation (1997). *Disaster preparedness for persons with disabilities: Improving California's response*. Retrieved 18 October 2005, from the World Wide Web: http://www.atnet.org/education/prepare.pdf

Campbell, J., & Oliver, M. (1996). *Disability politics: Understanding our past, changing our future*. London: Routledge.

Center for International Rehabilitation (2005). *Disability and early Tsunami relief efforts in India, Indonesia and Thailand.* Chicago: Center for International Rehabilitation.

Chang, C. M., Connor, K. M., Lai, T. J., Lee, L. C., & Davidson, J. R. T. (2005). Predictors of posttraumatic outcomes following the 1999 Taiwan earthquake. *Journal of Nervous and Mental Disease, 193*(1), 40-46.

Chari, R., & Padmanabhan, R. (2005). *Report on the visit to the Andaman & Nicobar islands to study the impact of the Tsunami tragedy vis-à-vis disability.* National Centre for Promotion of Employment for Disabled People/DPI India. Retrieved 15 November 2005, from the World Wide Web: http://ncpedp.org/policy/pol-res01.htm

Charlton, J. (1998). *Nothing about us without us: Disability oppression and empowerment.* Berkeley: University of California Press.

Chou, Y. J., Huang, N., Lee, C. H., Tsai, S. L., Chen, L. S., & Chang, H. J. (2004). Who is at risk of death in an earthquake? *American Journal of Epidemiology, 160*(7), 688-695.

DP2 (2001). *Disabled people and disaster planning.* Retrieved October 2005, from the World Wide Web: http://www.citycent.com/dp2/

Driedger, D. (1989). *The last civil rights movement.* London: Hurst & Co.

Duyan, V., & Karatas, K. (2005). Effects of the 1999 earthquake on the completely blind living in and outside Marmara, Turkey. *International Social Work, 48*(5), 609-619.

Fleischer, D. Z., & Zames, F. (2001). *The disability rights movement: From charity to confrontation.* Philadelphia, PA: Temple University Press.

Frampton, S., McNaught, A., Hardwick, J., & Chaffey, J. (2000). *Natural hazards: Causes, consequences and management* (2nd ed.). London: Hodder and Stoughton.

Gignac, M. A. M., Cott, C. A., & Badley, E. M. (2003). Living with a chronic disabling illness and then some: Data from the 1998 ice storm. *Canadian Journal on Aging–Revue Canadienne Du Vieillissement, 22*(3), 249-259.

Handicap International (2005a). *Tsunami in South East Asia: Handicap International's work in India.* Retrieved November 2005, from the World Wide Web: http://www.handicap-international.org.uk/page_479.php

Handicap International (2005b). *Tsunami in South East Asia: Handicap International's work in Sri Lanka.* Retrieved November 2005, from the World Wide Web: http://www.handicap-international.org.uk/page_52.php

Handicap International (2005c). *Tsunami: one year on...* Woking: Handicap International.

Hastie, R. (1997). *Disabled children in a society at war: A casebook from Bosnia.* Oxford: Oxfam.

Hemingway, L., & Priestley, M. (2006, forthcoming). Natural hazards, human vulnerability and disabling societies: A disaster for disabled people? *Review of Disability Studies.*

Hilhorst, D. (2003). Responding to disasters: Diversity of bureaucrats, technocrats and local people. *International Journal of Mass Emergencies and Disasters, 21*(1), 37-55.

Human Rights Watch (2005). *After the deluge: India's reconstruction following the 2004 Tsunami.* Retrieved from the World Wide Web: http://hrw.org/reports/2005/india0505/

Ingstad, B., & Reynolds Whyte, S. E. (1995). *Disability and culture*. Berkeley: University of California Press.

Interagency Network for Education in Emergencies (INEE) (2005a). *Inclusive education of children at risk*. Retrieved November 2005, from the World Wide Web: http://www.ineesite.org/inclusion/disabled.asp

Interagency Network for Education in Emergencies(INEE) (2005b). *Rebuilding for inclusiveness*. Retrieved November 2005, from the World Wide Web: http://www.ineesite.org/inclusion/default.asp

International Federation of Red Cross and Red Crescent Societies (IFRCRCS) (1994). *Code of conduct for The International Red Cross and Red Crescent movement and NGOs in disaster response programmes*. Retrieved September 2005, from the World Wide Web: http://www.ifrc.org/publicat/conduct/code.asp

International Strategy for Disaster Reduction (1994). *Yokohama strategy and plan of action for a safer world*. Retrieved November 2005, from the World Wide Web: http://www.unisdr.org/eng/about_isdr/bd-yokohama-strat-eng.htm

Keen, D. (2001). The political economy of war. In F. Stewart & V. Fitzgerald (Eds.), *War and underdevelopment: The economic and social consequences of conflict* (Vol. 1) (pp. 39-66). Oxford: Oxford University Press.

Kett, M., Stubbs, S., & Yeo, R. (2005). *Disability in conflict and emergency situations: Focus on Tsunami-affected areas*. Norwich: International Disability and Development Consortium/University of East Anglia.

Livanou, M., Kasvikis, Y., Basoglu, M., Mytskidou, P., Sotiropoulou, V., Spanea, E., Mitsopoulou, T., & Voutsa, N. (2005). Earthquake-related psychological distress and associated factors four years after the Parnitha earthquake in Greece. *European Psychiatry, 20*(2), 137-144.

Matsuoka, T., Yoshioka, T., Tanaka, H., Ninomiya, N., Oda, J., Sugimoto, H., & Yokota, J. (2002). Long-term physical outcome of patients who suffered crush syndrome after the 1995 Hanshin-Awaji earthquake: Prognostic indicators in retrospect. *Journal of Trauma-Injury Infection and Critical Care, 52*(1), 33-39.

Naghi, T. M., Kambiz, K., Shahriar, J. M., Afshin, T., Reza, S. K., Behnam, P., & Bahador, A. H. (2005). Musculoskeletal injuries associated with earthquake–A report of injuries of Iran's December 26, 2003 Bam earthquake casualties managed in tertiary referral centers. *Injury-International Journal of the Care of the Injured, 36*(1), 27-32.

Nakagawa, Y., & Shaw, R. (2004). Social capital: A missing link to disaster recovery. *International Journal of Mass Emergency and Disasters, 22*(1), 5-34.

National Organization on Disability (NOD) (2004). *Emergency preparedness survey: Final report*. Washington, DC: NOD.

National Organization on Disability (NOD) (2005). *Report on special needs assessment for Katrina evacuees (SNAKE) Project*. Washington, DC: NOD.

Oliver, M. (1983). *Social work with disabled people*. Basingstoke: Macmillan.

Oliver, M. (1990). *The politics of disablement*. Basingstoke: Macmillan.

Oliver, M. (1996). *Understanding disability: From theory to practice*. Basingstoke: Macmillan.

Oosters, B. (2005). *Looking with a disability lens at the disaster caused by the Tsunami in South-East Asia*. Brussels: CBM International.

Osaki, Y., & Minowa, M. (2001). Factors associated with earthquake deaths in the Great Hanshin-Awaji earthquake, 1995. *American Journal of Epidemiology, 153*(2), 153-156.

Perrin, P. (1998). *The impact of humanitarian aid on conflict development.* Retrieved December 2005, from the World Wide Web: http://www.icrc.org/Web/Eng/siteeng0.nsf/iwpList167/FA943455FADB895EC1256B66005C0A26#

Priestley, M. (2005). Disability and social inequality. In E. Margolis (Ed.), *Blackwell companion to social inequalities* (pp. 372-395) . Oxford: Blackwell.

Quinn, G., & Degener, T. (2002). A survey of international, comparative and regional disability law reform. In M. Breslin & S. Yee (Eds.), *Disability rights law and policy: International and national perspectives* (pp. 3-129). New York: Transnational.

Shaw, R., & Goda, K. (2004). From disaster to sustainable community planning and development: The Kobe experiences. *Disaster Prevention and Management, 28*(1), 16-40.

Smith, K. (2001). *Environmental hazards: Assessing risk and reducing disaster* (3rd ed.). London: Routledge.

Sphere Project (2004). *Humanitarian charter and minimum standards in disaster response.* Geneva: Sphere.

Sustainable Environment and Ecological Development Society (2001). *High powered committee on disaster management: Report on women, children, aged and challenged in disasters.* Government of India. Retrieved 10 October 2005, from the World Wide Web: http://www.benfieldhrc.org/disaster_studies/disability&disasters/hpc_seed.pdf

Swain, J., French, S., Barnes, C., & Thomas, C. (Eds.) (2004). *Disabling barriers–Enabling environments.* London: SAGE.

Takahashi, A., Watanabe, K., Oshima, M., Shimada, H., & Ozawa, A. (1997). The effect of the disaster caused by the great Hanshin earthquake on people with intellectual disability. *Journal of Intellectual Disability Research, 41*, 193-196.

Tierney, K., Petak, W., & Hahn, H. (1988). *Disabled persons and earthquake hazards.* Boulder: University of Colorado Institute of Behavioural Science, Program on Environment and Behaviour.

United Nations (2005). *Report of the World Conference on Disaster Reduction, Kobe, Hyogo, Japan, 18-22 January 2005.* Retrieved November 2005, from the World Wide Web: http://www.unisdr.org/wcdr/intergover/official-doc/L-docs/Final-report-conference.pdf

United Nations Commission on Human Rights (1998). *Guiding principles on internal displacement.* Retrieved December 2005, from the World Wide Web: http://www.unhchr.ch/Huridocda/Huridoca.nsf/0/d2e008c61b70263ec125661e0036f36e

US Geological Survey (2005). *Sumatra-Andaman Islands Earthquake.* Retrieved November 2005, from the World Wide Web: http://earthquake.usgs.gov/eqinthenews/2004/usslav/

USAID (1997). *USAID disability policy paper.* Washington, DC: US Agency for International Development Policy Paper PC09/12/97.

USAID (2004). *Policy on standards for accessibility for the disabled in USAID-financed construction.* US Agency for International Development. Retrieved October 2005, from the World Wide Web: http://www.usaid.gov/about/disability/financed_construction.html

Wisner, B. (in press). Disability and disaster: Victimhood and agency in earthquake risk reduction. In C. Rodrigue & E. Rovai (Eds.), *Earthquakes*. London: Routledge. Author manuscript accessed March 2006, http://www.radixonline.org/resources/disability_and_disaster_wisner.doc

World Health Organisation. (1999). *Community emergency preparedness: A manual for managers and policy-makers*. Retrieved November 2005, from the World Wide Web: http://whqlibdoc.who.int/publications/9241545194.pdf

World Health Organisation (2005). *Disasters, disability and rehabilitation*. Retrieved 8 October 2005, from the World Wide Web: http://www.who.int/entity/violence_injury_prevention/other_injury/disaster_disability2.pdf

Yeo, R. (2005). *After the Tsunami: Are disabled people being ignored?* ID21. Retrieved October 2005, from the World Wide Web: http://www.id21.org/insights/insights55/insights-iss55-art08.html

Zola, I. (1989). Towards a necessary universalizing of disability policy. *Millbank Memorial Quarterly, 67*(2), 401-428.

doi:10.1300/J198v05n03_02

Impact of the Indian Ocean Tsunami on the Well-Being of Children

Ngoh Tiong Tan

SUMMARY. This article provides an overview of the plight of the "tsunami generation" detailing the social and physical conditions after the disaster. Children are most vulnerable and need protection and clear and sustained support for rehabilitation. The social psychological recovery and protection of children are key concerns of child welfare and internal social service organisations. Policy and programmes should build on the people's resilience and provide the psychosocial and community support. doi:10.1300/J198v05n03_03 *[Article copies available for a fee from The Haworth Document Delivery Service: 1-800-HAWORTH. E-mail address: <docdelivery@haworthpress.com> Website: <http://www.HaworthPress.com> © 2006 by The Haworth Press, Inc. All rights reserved.]*

KEYWORDS. Children, tsunami, social recovery, rehabilitation, child protection, child welfare

---

Ngoh Tiong Tan, PhD, is Professor, Social Work Department, Augsburg College, MN. He was Associate Professor, Department of Social Work, National University of Singapore, and Chairman of FAST Project, and Immediate Past Regional President of International Federation of Social Workers, Asia Pacific.

Address correspondence to: Ngoh Tiong Tan, Social Work Department, Augsburg College, Campus Box #51, 2211 Riverside Avenue, Minneapolis, MN 55454 (E-mail: tan@augsburg.edu).

The author wishes to thank Benjamin Ng and Nicole Chew for their research assistance in the writing of this paper.

[Haworth co-indexing entry note]: "Impact of the Indian Ocean Tsunami on the Well-Being of Children." Tan, Ngoh Tiong. Co-published simultaneously in *Journal of Social Work in Disability & Rehabilitation* (The Haworth Press, Inc.) Vol. 5, No. 3/4, 2006, pp. 43-56; and: *Asian Tsunami and Social Work Practice: Recovery and Rebuilding* (ed: Ngoh Tiong Tan, Allison Rowlands, and Francis K. O. Yuen) The Haworth Press, Inc., 2006, pp. 43-56. Single or multiple copies of this article are available for a fee from The Haworth Document Delivery Service [1-800-HAWORTH, 9:00 a.m. - 5:00 p.m. (EST). E-mail address: docdelivery@haworthpress.com].

Available online at http://jswdr.haworthpress.com

doi:10.1300/J198v05n03_03

The Indian Ocean tsunami of Dec 26th, 2004 has impacted millions of lives. The strongest earthquake in 40 years, and the most powerful since the 20th century, it unleashed other tsunamis which caused great destruction on the coasts of Malaysia, Thailand, Sri Lanka, India, and the Maldives, crashing into their coastal lands, fishing villages and holiday resorts, wiping out neighbourhoods in their entireties, leaving flattened buildings and trees in its wake and killing some 169,752 people (Maps of the World, 2004). As the afflicted regions of Asia struggle to cope with recovery from the tsunami, many young voices are left unattended. These young voices represent humanity's fragile hope for a better future. Children and women are thought to be most vulnerable in any catastrophe and this one is no exception.

The tsunami has left 13,000 children without parents in the region, and more than a million either orphaned, injured, lost or negatively affected (Central News Network [CNN], 2005a). In Aceh alone, 35,000 children are either homeless, orphaned or separated from their parents, while in Sri Lanka and Thailand, almost 200 children and 300 children respectively have lost either one or both parents. Children from other countries at various places for holiday were not spared as they too were stripped away from their families (Straits Times [ST], 2005). Children are possibly the most fragile victims in a disaster. They have to deal with emotional, physical and psychological trauma that even an adult has difficulty coping with. Amongst the international organizations, UNICEF (United Nations Children's Fund) is one of the leading organizations in relief efforts to help children adversely affected by disasters, providing protection to the affected children.

This article concentrates on this "tsunami generation" (coined by UNICEF, 2005b), recollecting how they were initially affected by the earthquake and tsunami, assessing how they are currently coping and analyzing the kind of support they now need, including that of child protection, health care, education and social recovery.

IMMEDIATE AFTERMATH

The young lives of the tsunami generation were not safe even after having survived the initial destruction of this natural disaster. The aftermath of the tsunami seems deadlier than the initial onslaught. At least one-third of the three to five million people in the tsunami-affected region facing fundamental health risks and challenges to their survival were children. Since they are smaller and physically weaker, children are more susceptible to diseases. Cramped refugee camps provided very poor living conditions which were hazardous to anyone's health. Up to 40 people were sleeping in a room or tent. Food and clean water were in short supply and poor sanitation fostered conditions for measles, cholera and diarrhea outbreaks. Some of the children were malnourished and dehydrated. Furthermore, there were lingering pools of stagnant water left by the tsunami which provided breeding grounds for mosquitoes, which could transmit malaria, dengue fever and other diseases. The contaminated waters were an unwholesome source of water for thirsty children. Local Indian governments in the provinces of Tamil and Nadu and Kerala then coordinated efforts with UNICEF and the WTO to distribute vitamin A and measles vaccinations to the children who were in the refugee camps (CNN, 2005a; Fox News, 2005; Straits Times, 2005). UNICEF also made available medicinal and shelter supplies, as well as water purification supplies, recreation and school kits to camps in Banda Aceh, Indonesia, ensuring that children who were situated in areas not accessible by roads were not forgotten (UNICEF, 2005b).

Children were reportedly in danger of falling prey to crocodiles and wild snakes, which were coming out as a result of the levelling of sea levels, and the presence of thousands of dead bodies (Express India, 2005). The tsunami generation also face other threats. Governments and other child welfare organisations such as UNICEF, Plan and the Australian CCF are also extremely wary of child-traffickers who took opportunity of the chaos to kidnap or trick children for child labour or sexual exploitation. Children were also allegedly sold illegally to richer neighbouring areas (BBC News, 2005; Plan, 2005). Weak law enforcement and corruption, according to UNICEF, made it even easier for the child smugglers to ship children out of the country (UNICEF, 2005a). To intensify matters, the worst-hit part of Indonesia, Aceh, is within close proximity of Medan and Batam, infamous for being used by child-trafficking gangs to smuggle children out of the country. As of the 5th of January, 2005, 400 or more children were reported as missing in Aceh. While the numbers were not official, and anecdotal stories were the only sources of reports, organisations were certain that children were already being kidnapped and smuggled out of the country (Dursin, 2005). At the same time, in Andhra Pradesh, India, strangers were taking the children away from the relief centres, and in Thailand, a Swedish boy was kidnapped from the hospital (SBS, 2005; BBC News, 2005).

There were already reports of adults impersonating foster parents and smuggling children into Malaysia from Indonesia. From the CNN website, an e-mail message received by a UNICEF worker in Malaysia offered 300 orphans for adoption, promising that "all paperwork will be taken care of" (ABC News Online, 2005, p. 1).

PSYCHOLOGICAL WELL-BEING

As for the less immediate, but still vital psychological effects, one finds that the tsunami had a distressing, mental toll on the children. As volunteer Sureshna Saha said, "It's horrible. It's always worse when it's children." These children, being used to regular routines, were especially vulnerable to psychosocial stress, and the disruption of their schedules (especially with catastrophes like these) had "devastated their world," quoting Michael Wasserman, a pediatrician (CNN, 2005a, p. 1). This resulted in them becoming psychologically disoriented.

Some of the psychological scars that might affect the children are, as reported by CNN (2005b), post-traumatic stress disorder or even a phobia of water. "I used to play near the waves all the time back home, but I don't want to see it now," S. Chiranjivi, 12, related days after tsunamis destroyed his home in India's Hut Bay Islands. "I can't forget that day" (CNN, 2005a, p. 1). WebMD reported that fifty percent of the children who were exposed to the tsunami and its effects were likely to suffer from post-traumatic stress disorder, compared to thirty percent of adults who would suffer from it if exposed (Straits Times, 2005). It has been reported that the children were too traumatized to speak, and could only stare into space or cry. Children who were the oldest in the family now faced the intense pressure of being the head of the household (Straits Times, 2005; CNN, 2005a). Edge of the Australian CCF told reporters that their "coping mechanism" had "shut down" (SBS, 2005).

Furthermore, the stories and images of the tsunami also had a psychological impact on those children outside the areas hit. Stories and images of death and destruction made children worry about their own safety and security, resulting in depression.

CURRENT SITUATION

The biggest issue at hand was the problem of sanitation. According to a UNICEF report, with overcrowded living conditions in relief camps and dependence on a few sanitary toilets, camp organizers are trying to maintain good hygiene standards but are not sure that they are succeeding. Some of the makeshift refugee camps such as the Zionpuram camp in Kanyakumari, were converted from buildings of other uses such as church complexes. Thus, the facilities of the buildings are unable to cope with the needs the emergency demands. As there are only three toilets for the nearly 2,500 people living there, many people are left defecating in the open. In Sri Lanka, Indonesia and India, 87,160 people, 200,000 people and 189,000 people respectively have access to toilet facilities, assuming 20 people share a latrine.

One of the most current dangers faced by the children is the sudden outbreak of polio in the region. The UNICEF (2005a) web site reported on 6th May 2005 that the Indonesian Health Ministry had confirmed that another child was paralyzed by polio. This made it a total of five children who were diagnosed with polio and nine other suspected cases of polio. Even though Indonesia's immunization rate is seventy percent, there are still isolated places in which children are unable to obtain vaccines.

More complications are arising as loopholes in the registration system are being discovered. UNICEF has realised that an increase in consistency of standards in the registering of orphans is needed for more thorough monitoring of the children (International Organization for Migration, 2005).

As people with good intentions rush to adopt the orphans left behind by the tsunami, more restrictions are being enforced. While previously there was no restriction on the religion of the foster parents, the Indonesian government has now set a new criterion for foster parents, that they must be Muslim as well. This was done in response to the Muslim community who reacted when they heard that a Christian organisation was trying to find Christian foster parents for the Muslim orphans. The community is against the children of Indonesia having any possibility of being converted to Christianity or any other religion (Washington Post, 2005).

Presently, the government and aid groups are still struggling against the child-trafficking situation, which is also one of UNICEF's top priorities. Regarding the prevention of child-trafficking, Vice President Yusuf Kalla has issued an order banning Acehnese children from leaving the tsunami-stricken province unless permitted by their parents or immediate family members. Also, as babies and children below 15 are usually the targets of child traffickers, the government has issued a regulation banning Acehnese children below the age of 16 from leaving Indonesia and has imposed a temporary ban on their adoption, which would be in effect until the status of the orphans becomes more definite (BBC News, 2005; Washington Post, 2005).

Such a move has been received with varying degrees of enthusiasm. UNICEF (2005b) official, Simon Ingram stated on the behalf of the organization that they were "pleased" with the government's move. However, some other child activists such as Seto Mulyadi, chairman of National Commission on Child Protection, criticized the order. His stand was that "Acehnese children should be allowed to leave the province because the place is not favorable to their mental and emotional development" (UNICEF, 2005b, p. 1). This in turn conflicts with the view of the Save the Children organization, who has stated that the children's best chances of emotional recovery were in environments in which they were familiar (Save the Children, 2005).

On a positive note, some of the children seem to be recovering from the trauma of the disaster. In Galle, Sri Lanka, the orphans of Ruhunu State Receiving Home play cheerfully, and even greet relief volunteers with joy and mirth. The children also understand that the supplies being given to them are donations from people all over the world, and are extremely grateful and appreciative (Reuter AlertNet Foundation, 2005).

INTERVENTION AND SUPPORT NEEDED

Presently, all efforts are on rebuilding the communities which were completely abolished by the tsunami. Organizations such as UNICEF, the World Health Organisation, the United Nations and countless other aid groups have taken action to help the victims, collecting monetary donations to finance the reconstruction of the devastated areas. In a show of global unity, the United States, India, Japan, Australia and the United Nations committed themselves to form an international partnership to coordinate worldwide relief (Fox News, 2005).

UNICEF is currently providing clean water and sanitation to the afflicted region through various programs with the countries involved. Other programs involving the provision of clean water are also underway or in the midst of being carried out. Children are being vaccinated from fast-spreading diseases like measles, tetanus and gangrene. Rice, milk and other basic rations are provided. Tents are set up as temporary shelters for those that are displaced. Supported by UNICEF, Indonesian health officials are vaccinating children as quickly as they can to try to prevent further spread of diseases. More than 4,000 children in Cidahu, Indonesia, were immunized in a short span of two days, and over 1,300 children from the neighboring Bojong Genteng subdistrict were subsequently immunized as well. Furthermore, as of 1st June 2005, UNICEF began a massive $4 million polio immunization campaign on the island of Java, Indonesia, aiming to protect more than 6.4 million children from the crippling disease. Save the Children is providing medical care to stop outbreaks of diarrhea and other diseases which are deadly if not prevented (Wiseman, 2005). The need may be great, but the help received from all over the world matches the size of the need. Up to $4 billion worth of aid was raised from government, big international agencies and many tiny local groups (Haiken, 2005).

Another form of support for the orphans is to provide them with homes and families. The government and aid groups are presently trying to reunite the orphans with their next of kin, making sure that the people who claim the children are authentic relatives and not child traffickers. All separated children are especially vulnerable to abuse, exploitations and diseases. In an effort to reunite children with their families, aid agencies have set up a mass registration effort. This includes photographing all the children in the camps and temporary shelters and recording their identification details. This is done in hopes of matching children to their families (Haiken, 2005). In Sri Lanka, the government and child-protection activists are visiting homeless camps to register children to take them into official homes until the families come for them (Wiseman, 2005). All these measures were taken to counter child exploitation and trafficking. The Indonesian Government has intervened by issuing an edict that bars people under the age of 16 from leaving the country with children (Haiken, 2005).

As for reconstructive efforts, many organizations have also set up orphanages or children's homes to house the orphans. For example, Manna International, a help organization in India, is presently embarking on long-term relief. With one of their aims being to take care of over 100 orphans, they have opened one children's home and are set to open another in Andhra Pradesh (Manna International, 2005).

All the child experts and the child welfare organisations have agreed that it is crucial to get all the children back to school and to set up regular routines. By this, they would at least be able to find a sense of normalcy (CNN, 2005b). A sense of safety and security is very important at this time. Safety measures and a set routine introduced provide for better control and predictability. As such, orphanages and temporary schools have opened. Donations of school books, uniforms, shoes, desks and chairs have also been distributed (Reuter AlertNet Foundation, 2005). As school buildings have been crushed by the tsunami, improvisation of school buildings has been made using tents. UNICEF and the International Organisation for Migration are currently undergoing an accelerated reconstruction project in which 200 temporary primary schools in Aceh's most affected areas have been built in time for Indonesia's school year which began in July 2005. At the same time, new permanent school buildings as well as repairs to the old school buildings would continue to be made (United Nations Office of the Humanitarian Coordinator for Indonesia, 2005; International Organization for Migration, 2005).

However, since going to school may be too expensive as the children need to buy school books, uniforms and shoes, Indonesia's most successful fashion designer Harry Darsono has offered his services to train the orphans vocationally, so that they can learn to feed and clothe themselves. Under his expertise, the orphans may choose to learn sewing skills such as embroidery and needlework which Northern Sumatra is famous for instead of going to school (Zwire, 2005).

Five months into the disaster, with the monetary and practical help of world organisations, other help groups and even individuals and schools from other countries, reconstruction efforts are in place to rebuild communities and neighborhoods so that life can go on. With the physical needs of the orphans having been more or less met, what is now imperative is that the children get adequate emotional and psychological support. That the children recover well from the trauma is extremely crucial, as carrying emotional and psychological scars can torment them for the rest of their lives, and prevent them from ever leading normal or peaceful lives. Addressing and healing the children's psychological scars has to be done as soon as possible in order to have as little damage to the children's psychological welfare as possible (CNN, 2005b). The creation of "Child-Friendly Spaces" (CFS) in camps and communities provides a platform from which children learn age-relevant competencies that help them cope with the risks and effects. In such an area where children meet other children, a sense of safety, structure and continuity enables them to find psychosocial and physical support encouraging healthy development. Such an approach would assess the need of children for special attention amongst those that are recovering well from the trauma. It is a place where children feel comfortable and confident enough to talk about issues like abuses (Curtis, 2005).

As stated previously, it is vital that school continues for the youngsters, even if there is no formal curriculum. With relative normalcy, children can begin to recover psychologically. It is also in school where teachers are able to keep an eye out for children who are not coping well psychologically (UNICEF, 2005b), and help them recover from the trauma (Yahoo UK News, 2005). As such, teachers and volunteers in the schools, homes and orphanages have to be extremely alert to identify the ones who are not adjusting well, and discerning enough to see to the children's emotional needs. The teachers and volunteers coming from other countries also need to be trained to be sensitive to the different cultures, traditions and ways of life that the children are used to. This issue has been attended to by aid groups and experts who are presently training teachers for just this purpose. They are also assisting in setting up schools and establishing other social activities (CNN, 2005b). These activities include drama, music and sports which allow the children to safely express their feelings and cope with their situation (Save the Children, 2005). A Buddhist temple had been transformed into a refugee camp where boys can play cricket, using coconut husks and tree branches. Likewise, a Muslim college lets children play on its playgrounds. The Christian Children's Fund is also clearing out spaces in order that children can learn and play, hoping to encourage them to express their emotions through songs and craft. This has proven to have therapeutic effects for the children affected by the tsunami (Wiseman, 2005).

Regarding issues of adoption, Makmur Sunusi, Director for Child Protection at the Ministry of Social Affairs of Indonesia, said that adoption is possible only when it can be verified that both parents and other immediate relatives of the child have been killed in the tsunami disaster. At that point, the teachers, schools and orphanage then have to prepare the remaining orphans for the idea of adoption (Straits Times, 2005). This task has to be done carefully and conscientiously, as there could be dire consequences of quickly placing a child with foster parents. Situations such as relatives turning up to claim an orphan who had already adjusted well in his/her foster home or orphans being sexually abused or made to do forced labor by their foster parents could occur (CNN, 2005c).

Yet adoption in this emergency phase, especially inter-country ones, seems to be the last resort to the adoption experts and the Indonesian government. Their opinion is that in the best interests of the orphans, people should donate money to causes directly helping the children such as helping to rebuild their community and schools, and not adopt them, as taking traumatized children away from their familiar environment would only serve to traumatize them further (CBC News, 2005). The Social Affairs Ministry in Indonesia issued a statement saying that adoption would only be used as a last resort, as the government prefers using family-based care (Reuters, 2005). Australia supports the position taken by UNICEF and the Indonesian authorities. Hence, Australia is supporting international agencies to assist children in reestablishing the lives that were lost in the waves. Recognizing the need of supporting national efforts, Australia's strategy is to direct money to agencies with a mandate to deal with child protection (AusAid, 2005).

A number of people in Jakarta have expressed their desire to become foster parents to Acehnese orphans. The Council of Indonesian Ulemas (MUI) has made it known that parents who want to adopt should be economically stable and have the same religion as the child. As such, another possible intervention could be helping screen and assess the suitability of the foster parent as it might help bring some of these young tsunami survivors to a better life.

A study done six months after the disaster evaluated the effectiveness of interventions in terms of the distribution of supplies and aid. The research team found differences in the distribution of help provided by both government and non-governmental agencies to survivors. This is the result of alleged corruption, favoritism and poor management. Inconsistencies in help-giving were reported in India, Thailand and Sri Lanka. Some fishermen, for example, received aid to rebuild their boats while fishermen in nearby villages did not receive the same aid. Furthermore, the research shows that there is a lack of community participation during the reconstruction process. In Sri Lanka, some of the people rioted against the government. In Thailand, people criticized the authorities for giving the survivors a once-off payment rather than a long-term process of reconstructing the community. All these examples showed that no matter how much aid is available, the vulnerable populations are made even more vulnerable during such times. Historically marginalized groups suffer discrimination and government relief failed to meet their needs adequately (Deen, 2005).

Finally, it is important to note that part of the protection of children in emergency response situations involves the implementation and adherence to a child protection policy. Such a policy explicitly provides for the safety of children through the commitment of minors to a child-safe organization. The interventions seek to be inclusive and integrative by providing support that is within communities. It is common that the most vulnerable are left out during interventions. Hence, it is crucial to seek out those that are not in the mainstream and to give them the appropriate help that they need (Curtis, 2005).

SOCIAL WORK RESPONSE TO DISASTER

International Federation of Social Workers (IFSW) and Commonwealth Organization for Social Work (COSW), as a response to the Tsunami disaster, have established the FAST (Families and Children of Tsunami) Project (2006).

The goals of the FAST Project are:

1. To meet the social and emotional needs of children, families and survivors of the Tsunami; and
2. To ensure capacity building of social workers, local NGOs and communities affected by the disaster.

The first goal is achieved through the second.

The process of social work is anticipated to be limited involving further assessment and planning on the ground, collaboration with local social workers on specific projects to ensure cultural sensitivity and relevance, development of personnel and support, and implementation and evaluation of the project. The focus of the project is on the needs of the children, youth and families in terms of their emotional adjustment, and grief issues for the entire communities. Urgent attention is needed for individual emotional support and mental health services to those immediately impacted. Included in this is the management of the community's loss, facilitating its expression, and ensuring that accurate information is available and appropriate for all affected groups on coping and services available. Both trained staff and volunteers are empowered to provide this vital support.

The strategy, thus, is to provide social-emotional support, advocacy and planning, and capacity building for family and community recovery. The key approach aims to support the community-level long-term recovery and rebuilding as well as to support local workers in reaching out to and assisting vulnerable groups. Currently, social workers as individuals and as a group (including social workers in various NGOs and international bodies and social work students) have been involved with direct work with the various tsunami-affected communities. The project initiated by the Sri Lanka Association of Professional Social Workers is directed at the psychosocial recovery of residents, especially the children in the district of Hambantota (Jayasinghe, 2006).

The call to raise support to enable local professional social work actions has met with some success in terms of providing financial grants for social work initiatives. More importantly, FAST has, through its website and networking, raised the level of awareness and solidarity in providing a platform for social work initiatives. The website has also provided information on projects, education and on-line articles of the tsunami, trauma and social work response. A key project actively implemented by the Indonesian Association of Professional Social Workers is the training of trainers, front-line care workers in Aceh and the proposal to develop a disaster management, trauma research and resource center in Indonesia and elsewhere. The FAST project thus plays a catalytic role in empowering social workers and social work organizations in the different tsunami-affected countries to take leadership and enhance their contribution and effectiveness in the community (FAST Project, 2006).

As professional social workers, the approach is not only on the relief work, but on the long-term impact of disaster on the people and the communities. It is not just the healing of children and people but also of communities and nations. The FAST project's contribution is in the areas of assessment and planning, which can only be effected through the creation of collaborative networks with social work colleagues, government and non-government organizations.

CONCLUSION

These children and orphans, the tsunami generation, are arguably the most hard-hit victims of this natural disaster, and the responses and efforts to help are extremely commendable. However, as easy as it is to concentrate relief efforts on getting these orphans physically comfortable, perhaps even more important is to ensure that the children recover psychosocially. The steps may be tenuous, and much patience and determination is indeed needed. However, Save the Children (2005) firmly believes that these young children, with their resilience, would recover eventually, even with first-hand experience of the tsunami.

In addressing the needs and resilience of children in tsunami and earthquakes or any disaster, the needs of individuals, families and communities in promoting capacity building and long-term development must be emphasized. With appropriate social support, and the care of familiar people in a familiar environment, the chances of the orphans recovering psychosocially is high. Perhaps, then, one can take hope from that, and persevere (Save the Children, 2005).

REFERENCES

ABC News Online (2005). *Concern for 'tsunami generation' increases*, Retrieved 1st June 2005 from the ABC News Online Website: http://www.abc.net.au/news/newsitems/200501/s1276893.htm

AusAid (2005). *Child protection*. Retrieved on 23 February from the AusAid Website: http://www.ausaid.gov.au/hottopics/tsunami/child.cfm

BBC News (2005). *Aid plea for 'tsunami generation.'* Retrieved 1st June 2005 from the BBC News Website: http://news.bbc.co.uk/1/hi/world/asia-pacific/4147669.stm

CBC News (2005). *Tsunami orphans*. Retrieved 1st June 2005 from the CBC News Website: http://www.cbc.ca/news/background/asia earthquake/tsunami-orphans.html

Central News Network (CNN) (2005a).*The most vulnerable victims*. Retrieved 1st June 2005 from the CNN Website: http://edition.cnn.com/2005/WORLD/asiapcf/01/05/overview.children/

Central News Network (CNN) (2005b). *Treating children's emotional wounds*. Retrieved 2nd June 2005 from the CNN Website: http://edition.cnn.com/2005/WORLD/asiapcf/01/05/tsunami.children.cope/index.html

Central News Network (CNN) (2005c). *Traffickers threaten Aceh orphans*. Retrieved 2nd June 2005 from the CNN Website: http://edition.cnn.com/2005/WORLD/asiapcf/01/04/indonesia.children/index.html

Curtis, P. (2005). *Child protection in emergencies–The Asian Tsunami*. Child Rights Policy Advisor, UK: World Vision.

Deen, T. (2005). *Tsunami impact: Women, children suffer most in overcrowded camps*. Retrieved on 23 February, 2005 from the Website: http://www.ipsnews.net

Dursin, R. (2005). *Aceh children becoming vulnerable to traffickers*. Retrieved 2 June, 2005 from the Inter Press Service News Agency Website: http://ipsnews.net/news.asp?idnews=26911

Express India (2005). *Stranded Indians battle snakes, crocodiles*. Retrieved 2nd June, 2005 from the Express India Website: http:/www.expressindia.com/fullstory.php?newsid=40494

FAST Project (2006). *About FAST*. Retrieved on 1st March, 2006: http://www.fastproject.net

Fox News (2005). *Tsunami death toll tops 119,000*. Retrieved 2nd June 2005 from the Fox News Website: http://www.foxnews.com/story/0,2933,142911,00.html

Haiken, M. (2005). *Saving the 'Tsunami generation': The challenges children face, and how you can help them*. Retrieved on 23 February, 2005 from the Baby Center Website: http://www.babycenter.com/refcap/toddler/toddlerfamily/1415288.html

International Organization for Migration (2005). *Indonesia disaster recovery program: 17-31 May 2005*. Retrieved 2nd June, 2005 from the ReliefNet Website: http://www.reliefweb.int/rw/RWB.NSF/db900SID/EVOD-6CWGQY?OpenDocument

Jayasinghe, R. (2006). *Progress report: Psychosocial care for Tsunami victims in Hambantota District*. Sri Lanka: Sri Lanka Association of Professional Social Workers.

Manna International (2005). *Tsunami relief*. Retrieved 2nd June, 2005 from the Manna International Website: http://www.manna7.org/tsunami.htm

Maps of the World (2004). *Earthquake, Tsunami kills thousands in South Asia on 26th December 2004*. Retrieved 2nd June, 2005 from the Maps of the World international Website: http://www.mapsofworld.com/world-news/26-12-2004-indonesia.html

Plan. (2005). *Child protection post tsunami*. Retrieved 2nd June, 2005 from the Plan Website: http://www.plan-uk.org/action/emergencies/tsunamidisaster/protectingchildren posttsunami/?view=textonly

Reuter AlertNet Foundation (2005). *U.N. fears for 'tsunami generation' of children*. Retrieved 2nd June, 2005 from the Reuter AlertNet Foundation Website: http://www.alertnet.org/thenews/newsdesk/L04718616.htm

Reuters (2005). *Up to 10,000 Aceh children seek parents*. Retrieved 2nd June, 2005 from the MSNBC News Website: http://www.msnbc.msn.com/id/6946276/

Save the Children (2005). *Policy Brief,* 1(1), 2-12.

SBS (2005). *'TSUNAMI GENERATION' tries to cope*. Retrieved 1st June, 2005 from the SBS Website: http://www9.sbs.com.au/theworldnews/region.php?id=102563&region=2

Straits Times (ST) (2005). *Tsunami orphans' grim future*. Retrieved 1st June, 2005 from The Straits Times Website: http://straitstimes.asia1.com.sg/sub/topstories/story/0,5562,294102,00.html?

United Nations Children's Fund (UNICEF) (2005a). *At a glance Indonesia*. Retrieved 2nd June, 2005 from the UNICEF Website: http://www.unicef.org/infobycountry/indonesia.html

United Nations Children's Fund (UNICEF) (2005b). *UNICEF identifies key steps to save 'Tsunami Generation.'* Retrieved 1st June, 2005 from the UNICEF Website: http://www.unicef.org/media/media_24721.html

United Nations Office of the Humanitarian Coordinator for Indonesia (2005). *Indonesia Earthquakes Field Situation Report*. Retrieved 1st June, 2005: Relief Web, http://www.reliefweb.int/rw/rwb.nsf/db900SID/DDAD-6B2MWE?OpenDocument

Washington Post (2005). *Tsunami orphans won't be sent to Christian home*. Retrieved on 1st October, 2005: www.washingtonpost.com/wp-dyn/articles/A7535-2005Jan13.html

Wiseman, P. (2005). *Children watch as waves sweep away all they know*. Retrieved on 23rd February, 2005 from the USA Today Website: http://222.usatoday.com/news/world2005-01-09-tsunamikids-cover_x.htm

Yahoo UK News (2005). *U.N. fears for 'tsunami generation.'* Retrieved 1st June, 2005 from the Yahoo News website: http://uk.news.yahoo.com /050104/325/f9nap.html

Zwire (2005). *Vocational training for kids urged after tsunami*. Retrieved 2nd June, 2005 from the Zwire Website: http://www.zwire.com/site/news.cfm?newsid=14616402&BRD=1817&PAG=461&dept_id=222087&rfi=6

doi:10.1300/J198v05n03_03

The State and Civil Society in Disaster Response: Post-Tsunami Experiences in Tamil Nadu

Krithika Srinivasan
Vijay K. Nagaraj

SUMMARY. This paper, based on the report "The State and Civil Society in Disaster Response: An Analysis of the Tamil Nadu Tsunami Experience," (Srinivasan, Nagaraj, & Venkatesh, 2005) is essentially an empirical analysis of state and civil society responses in Tamil Nadu (India) to the tsunami of December 26, 2004. It examines interventions by state and non-state agencies, as well as people's experiences in the relief and rehabilitation phases to identify factors influencing both positive and negative outcomes of the tsunami response. Issues related to vulnerability and exclusion, equity, transparency and accountability in different sectors of disaster intervention are explored to highlight themes revolving around reach and efficacy of relief and recovery processes. These analyses bring out some interesting lessons with regard to the importance of institutional autonomy, non-politicized decision-making, and synergetic state-civil society interfaces in fostering inclusive, transparent and accountable rehabilitation processes. The roles played by in-

Krithika Srinivasan, MSW, is Program Associate, Tata Institute of Social Sciences, Mumbai, India (E-mail: krithikasrinivasan@yahoo.com).

Vijay K. Nagaraj, MSW, is Faculty Member, Director's Office, Tata Institute of Social Sciences, Mumbai, India (Email: vijay@tiss.edu).

[Haworth co-indexing entry note]: "The State and Civil Society in Disaster Response: Post-Tsunami Experiences in Tamil Nadu." Srinivasan, Krithika, and Vijay K. Nagaraj. Co-published simultaneously in *Journal of Social Work in Disability & Rehabilitation* (The Haworth Press, Inc.) Vol. 5, No. 3/4, 2006, pp. 57-80; and: *Asian Tsunami and Social Work Practice: Recovery and Rebuilding* (ed: Ngoh Tiong Tan, Allison Rowlands, and Francis K. O. Yuen) The Haworth Press, Inc., 2006, pp. 57-80. Single or multiple copies of this article are available for a fee from The Haworth Document Delivery Service [1-800-HAWORTH, 9:00 a.m. - 5:00 p.m. (EST). E-mail address: docdelivery@haworthpress.com].

Available online at http://jswdr.haworthpress.com

doi:10.1300/J198v05n03_04

stitutional responsiveness and flexibility in shaping an effective disaster response also emerge very clearly from this study of the Tamil Nadu experience. Another crucial finding points to the need for detailed, reliable and disaggregated geo-demographic and socioeconomic records as a resource base for informing relief and rehabilitation interventions. The study draws extensively from the experiences and insights of people affected by, and involved in tsunami response, and from secondary knowledge resources available on the disaster. doi:10.1300/J198v05n03_04 *[Article copies available for a fee from The Haworth Document Delivery Service: 1-800-HAWORTH. E-mail address: <docdelivery@haworthpress.com> Website: <http://www.HaworthPress.com>*

KEYWORDS. Tamil Nadu, tsunami, civil society response, state response, accountability, participation information

It was a devastating earthquake off the coast of Sumatra in the Indonesian archipelago, of magnitude 9.0 on the Richter Scale, followed by one of magnitude 7.3 on the Richter Scale 81 kilometers off Pulo Kunji, Great Nicobar, India, that resulted in massive tsunamis in several countries in South Asia and East Africa–Indonesia, Sri Lanka, India, Thailand, Somalia, Myanmar, Maldives, Malaysia, Tanzania, Bangladesh, Kenya and Seychelles.

In India, the tsunami affected nearly 2,260 kilometers of the mainland coastline (Tamil Nadu, Kerala, Andhra Pradesh and Pondicherry), as well as the Andaman and Nicobar Islands, with tidal waves up to 10 meters high penetrating up to 3 kilometers inland, taking at least 10,749 lives (Government of India, 2005) and affecting more than 2.79 million people across 1,089 villages.

In Tamil Nadu, the worst affected state on the mainland, over 8,010 deaths were reported. Thirteen districts were hit by the tsunami, of which Nagapattinam, Kanyakumari, and Cuddalore were the worst affected, with a death toll of 6,065,828, and 617, respectively. In all, according to the GoTN (Government of Tamil Nadu), 984,564 people were affected with over 8,000 deaths and 126,182 homes being damaged or destroyed. In addition, more than 3,400 people were reported missing ("Tsunami–The Killer waves," 2005).

In Tamil Nadu, one thousand kilometers (kms) of the coastline were affected, with water penetrating up to 1-1.5 kms inland and causing extensive damage to nearly 19,168 hectares of cropped area (damage figures, however, have been constantly evolving). In addition, Tamil Nadu also suffered substantial losses in terms of loss of livelihoods–loss of assets such as boats and other fishery related assets, loss of livestock, disruption of the fisheries sector, traditional market links and other livelihoods linked to the coastal economy, and damage to other resources such as pasture and grazing lands. The tsunami also caused extensive damage to social infrastructure, community assets, harbors, and transport, power and communication infrastructure.

This paper is based on a study conducted in four tsunami affected districts of Tamil Nadu–Chennai, Kanyakumari, Nagapattinam and Cuddalore–nearly a year after the tsunami. The primary objective of the study, and of this paper, is to analyze and bring together the experiences and learnings of a range of actors affected by, and involved in, the disaster response in order to identify factors that influence effectiveness and reach of the relief and rehabilitation processes. These analyses include examinations of the state and civil society responses, key issues relating to the relief and rehabilitation processes, and transparency and accountability concerns. The paper relies on fieldwork, discussions with key actors in the tsunami response, and also draws from the wealth of material available from various agencies on the tsunami.

THE STATE RESPONSE

At the very outset, it is important to stress that, as a phenomenon, the tsunami was one that was beyond the pale of experience of the State apparatus. Despite an initial paralysis due to the nature and sheer scale of the disaster, compounded by the breakdown of communications channels, the State and District level administrations (the latter is headed by the District Collector and Magistrate, a senior officer of the Indian Administrative Services) swung into action by noon 26 December 2004, with the first government orders (GOs) regarding relief and rehabilitation packages being issued on the 28th December 2004.

Bringing in the Armed Forces for search and rescue operations, and moving on to organize relief camps and announce ex-gratia payments for loss of life and injury, the government, in the relief phase, is to be noted for averting a large-scale public health crisis despite thousands being housed in relief camps, and extensive destruction of essential infrastructure. In the rehabilitation phase, the State's efforts have largely focused on housing and livelihood rehabilitation, with a massive reconstruction program well underway. Requirements of vulnerable groups such as children, destitute individuals and women have also been addressed to an extent through measures such as school fee waivers, special deposits, pensions, etc.

Critical Institutional Factors

To enhance the response capacity of the administration, a number of carefully chosen personnel from various departments across the State were put in place at various levels and also given considerable decision-making powers. In Nagapattinam, the worst affected district, 11 teams, each comprising one Indian Administrative Services (IAS) officer, senior officials from departments such as Health, Agriculture, and Public Works, and headed by a Minister of the State Government, were formed to assess damages, coordinate relief processes, and distribute ex-gratia in specific village clusters. Such teams, working on similar lines, were formed in other districts as well. These teams, which were in place for a month, were given administrative and financial powers to enable quick response to the situation on the ground. They also established linkages with civil society responders and, in all, have been recognized for the vital role they played in ensuring effective relief processes.

Across the state, postings of senior officers to the affected districts were executed rapidly, with personnel being chosen for their competency, commitment, integrity, and familiarity with local conditions and affected communities. The State Government promptly devolved financial and administrative powers, with District Collectors being authorized to draw up to 10 million rupees to deal with a range of immediate requirements according to their discretion. The administrative and financial devolution was extended to lower levels also (such as officers responsible for relief camps), facilitating rapid and locally adapted decision-making. Further, while political representatives did accompany relief teams, there seem to have been clear guidelines issued from the top (the State Government) that led to minimal political interference in the relief process that was largely driven by the administration. It is stressed by many observers that the key to a strong institutional response after the initial paralysis lay in the above conditions being met.

This level of administrative autonomy and lack of political interference has continued even through the rehabilitation phase, and coupled with an efficient and less bureaucratic leadership handpicked to lead the tsunami response, is said to have made GoTN's response, despite several shortcomings, more effective as compared to other states, and according to one observer, even some other countries. Further, the Tamil Nadu experience shows that an efficient administration empowered with sufficient autonomy can actually compensate for shortcomings in the policy framework. Given the fact that policy frameworks are not always consistent with local conditions, district administrations have been very flexible in allowing necessary adaptations in implementation of rehabilitation programs. A case in point is the reconstruction of houses within the 200m CRZ zone in Kanyakumari (see section on housing rehabilitation). It is also recognized that it was a clear message from the strong political leadership recognizing the importance of autonomy and protecting the administration from political interference that was crucial in fostering this facilitative environment.

These institutional factors have translated into a high degree of government responsiveness to the situation on the ground. The administration has been extremely proactive in engaging with civil society in terms of partnering with a range of non-government agencies in the relief and rehabilitation processes, as well as in being extremely receptive to feedback regarding policy frameworks and efficacy of rehabilitation processes. Coordination mechanisms have been set up for liaison with civil society, and bureaucratic procedures cut down drastically. This responsiveness has also meant an enhanced people-administration interface: (a) group discussions in affected communities conducted by Collectors in the initial days; (b) regular field visits; (c) weekly grievance days; and (d) officers specially designated to handle particular aspects of relief and reconstruction to enable a disaggregated response to grievances; these officers' contact details are made public through the local media, etc.

It must be noted here that elected village *panchayats* (constitutionally recognized local self-governing bodies at the village level) have not been really visible and central in the response efforts. It is believed by many observers that these elected bodies have been bypassed, albeit quietly, in an attempt to keep the relief and rehabilitation process in the hands of the administration rather than vest it in elected, i.e., political bodies, and thus minimize the chances of politicizing the relief and reconstruction process. However, this move has been much questioned and debated on the grounds that these are democratic bodies, perhaps the best suited to implement an equitable rehabilitation process.

CIVIL SOCIETY RESPONSES

In the wake of the tsunami, civil society, in India and abroad, rose to the occasion, with communities, individuals, bilateral donor agencies, international, national and local non-governmental organizations (NGOs), academic institutions, political groups, the media, religious trusts, interest groups with membership bases in the affected communities, and the corporate sector mobilizing to respond to the needs of the tsunami-affected areas.

To begin with, the general public and civil society organizations brought in with them substantial amounts of resources and expertise to help provide immediate relief to the affected people, and were engaged in a whole spectrum of activities–rescue, retrieval and disposal of bodies and debris, provision of food and essential medical care, construction of temporary shelters, psychosocial interventions, setting up relief camps, etc. One of the major contributions of the civil society organizations has been their ability to fill crucial gaps at the time of relief and rescue. For instance, while the government could mobilize large quantities of food-grain, it was civil society organizations that were able to mobilize essentials such as children's clothing, underwear, sanitary pads and baby food.

Civil society engagement in tsunami response has continued into the rehabilitation phase, with non-government agencies extensively involved in not only planning, resourcing and implementing rehabilitation programs, but also in influencing state policy to shape it in a manner that addresses people's real needs. For instance, it was civil society that was instrumental in bringing to the government's attention that coastal communities comprised not only fish workers, but also a range of other groups involved in diverse occupations, which had all been affected in various ways by the tsunami.

Civil society organizations have been proactive in highlighting the specific and precise needs of communities through intervention, research and documentation, bringing to light instances of exclusion, and in advocating for the rights of vulnerable groups such as *Dalits* (literally meaning "broken people"; *Dalit*, like *Black*, is a political identity of a section of people formerly referred to as "untouchables" or "outcastes," i.e., those outside the "Caste" Hindu society), non-fish workers, women and the aged. They also have played crucial roles in giving feedback to the administration regarding shortcomings and malpractice in various rehabilitation processes, and in bringing to the government's attention any negative implications of state policies–for instance, older men were marrying minor orphaned girls for the special deposits made in their name by the government as a part of tsunami rehabilitation. At this juncture, it must be noted that the government, in general, has been proactive in taking corrective measures and passing the necessary directives.

In addition to organizations directly involved in the relief and reconstruction work, there are also a number of groups and collectives that focus entirely on independent analysis of policy and practice, using the media, the courts, and mobilizing affected people and communities to highlight key issues of concern. Given their non-involvement in the actual administration of relief and rehabilitation, they often act as watchdogs trying to examine the situation from wider perspectives that distance can often bring, but are, of course, unable to always appreciate the challenges faced while actually trying to implement policy or operationalize standards and plans.

THE STATE AND CIVIL SOCIETY: INTERFACES, SYNERGIES, AND CONFLICTS

The response to the tsunami has seen an extraordinary level of synergy between various actors–government and non-government. It can be stressed that coordination, dialogue and information sharing between civil society organizations and the government have gone a long way in enhancing the effectiveness and reach of the response.

Right from the beginning, the administration was open to working in tandem with a range of civil society groups. This was seen during the relief phase, with the specially appointed teams regularly meeting, and proactively welcoming feedback from civil society groups. In addition, the administration also played a crucial role in coordinating interventions by NGOs to avoid duplication, and often sought their (NGOs) help in filling gaps in the government response. Such close engagement has continued through the response processes, with civil society working closely with the state in direct interventions, as well as in providing policy related feedback and input. The government, on its part, has been extremely responsive to civil society's concerns, as is visible in the constant evolution of tsunami GOs based on field information from various civil society sources, including the media.

The entire state-civil society interface appears to be built around two key principles adopted by the government:

a. Accessibility–The highest levels of the administration at the District and State levels were very willing to meet, listen to and pay attention to the issues and concerns raised by civil society, as well as accept several of their recommendations.
b. Inclusiveness–The government has worked actively to create spaces and mechanisms for civil society organizations to participate effectively.

The Public-Private Partnership Framework

One such mechanism is the public-private partnership framework adopted by the government for the rehabilitation process. Under this framework, civil society organizations were invited to invest in and partner with the government in livelihood rehabilitation, and housing and infrastructure reconstruction.

This partnership has been widely adopted especially for housing reconstruction–government is responsible for providing land and basic infrastructure, while civil society organizations are to provide resources for, and undertake the actual construction process. State policy also lays down minimum standards for disaster resistant housing, guidelines relating to community participation, insurance, monitoring and third party auditing. If, for a particular habitation, no civil society organization comes through to undertake construction, then the affected people themselves would be permitted to do so with financial and technical assistance from the state. The government is to undertake construction only in cases in which both NGOs and beneficiaries are not forthcoming. In the urban areas of Chennai and Thiruvallur, however, all reconstruction is being undertaken by the government.

Coordination Forums

Coordination mechanisms have been established at the District level (and at the State level at a later stage) to better leverage the expertise of civil society and optimize the use of resources. Coordination bodies (such as the Kanyakumari Rehabilitation Resource Centre, NGO Coordination and Resource Centre [NCRC] in Nagapattinam, and the Chennai NGOs Coordination Cell) that bring together a wide range of civil society organizations and the administration were set up by the joint initiative of certain NGOs and the Collectorate. In other districts, too, coordination meetings are organized regularly.

The single most important feature of these coordination bodies is that they have a strong working relationship with the administration. They help ensure a constant interface between the administration and civil society groups (and between civil society groups themselves), and usually have the following objectives:

a. Coordinating rehabilitation efforts by civil society organizations and the government to maximize potentials and avoid duplication;
b. Sharing knowledge, perspectives, information and expertise;
c. Discussing rehabilitation issues and giving feedback to the government.

Effective coordination is not just a felt need of the civil society organizations involved, but also greatly assists the administration in ensuring more effective resource allocation and planning, and has the potential to greatly reduce duplication and wastage of resources. A very interesting aspect of the experience of the district level coordination mechanisms is the role played by the District Collector in shaping civil society interventions through these forums. If the experience in Tamil Nadu is any indication, for such coordination mechanisms to be effective, key personnel such as the District Collector must have sufficient knowledge of and understanding regarding the work and dynamics of civil society organizations as well as possess good facilitation and coordination skills.

One of the criticisms leveled at the coordination and interface forums is that they are democratic only in name and, in effect are controlled by the administration. It is pointed out that these coordination bodies have largely restricted themselves to logistical issues, and have not served as forums where substantive and policy impacting discussions occur. It is also pointed out that the presence of these coordination bodies has not enhanced the overall transparency of either the government or non-government agencies. However, given their role in facilitating better communication and maximizing potentials, there is clearly a need for these mechanisms not only to continue, but also become stronger, as well as expand the scope of their functioning to include transparency and accountability related concerns.

Coordination and advocacy networks of a different nature can also be found in Tamil Nadu. The Tsunami Relief and Rehabilitation Coordination (TRRC) is a forum that brings together a large number of local NGOs, social movements and affected peoples' organizations from across Tamil Nadu, mostly with a focus on strengthening advocacy and coordinating efforts to influence policy and practice in the relief, reconstruction and rehabilitation process. The TRRC focuses its efforts on mobilizing affected communities to ensure effective rights-based rehabilitation and development. Some of the organizations affiliated with the TRRC are also involved in the implementation of housing and livelihood rehabilitation. The TRRC also engages in policy overview, looking at relevance and standards, as well as transparent and democratic functioning of both government and non-government agencies.

Civil Society Interventions: Conflicts and Debates

The extent of networking and synergy visible between state and non-state actors in the tsunami response is not equally visible when it comes to relationships *between* various civil society actors. Given the diversity in institutional natures, orientation and perspectives, and the extent of financial, human and technical resources available, coordination across the spectrum of civil society organizations has posed several challenges.

At one end of the spectrum are organizations that invest heavily and are intensely involved in rehabilitation-related work, typically sharing a strong working relationship with the government, and often being affiliated with coordination cells such as the NCRC. At the other end of the spectrum are coordination cells like the TRRC, which pursue an activist agenda and have taken on a monitoring role, often not hesitating to openly criticize the work of the government as well as some of the larger agencies, development organizations and INGOs engaged in relief and rehabilitation work.

Conversations with INGOs and the larger development organizations clearly reveal that many of them view the TRRC and many of the associated organizations as "confrontationist" and difficult to work with, and that they perceive themselves as more "constructive" in their approach than those associated with the TRRC.

On the other hand, there seems to be a perception that coordination cells such as the NCRC are "pro-government" rather than pro-people, and that they are often coopted by the government. Apparently their investments and involvement in the rehabilitation processes create the need to maintain smooth relations with the government, and therefore, they refrain from taking up controversial issues that could antagonize the administration. While at the one hand this could indeed be true, people associated with these coordination cells point out that this perception probably stems from the fact that they, having direct access to the upper echelons of the administration, do not usually need to make a hue and cry about their stand, or take it to the media.

Despite the vast differences among them, these apparently conflicting networks of organizations are actually complementary to each other–in any rehabilitation process, both activist organizations and development organizations have roles to play. Despite this truism, efforts to get the two ends of the spectrum to work together have been rather futile, with diverse kinds of civil society organizations arrayed along various axes rather than forming a loop that feeds into and complements each other's work.

Coordination between various organizations involved in direct rehabilitation has also been lacking, largely due to issues related to competition, and an unwillingness to share information. These conflicts and difficulties in coordinating interventions have had negative implications in terms of accountability to people and an undermining of response efforts, and pose a challenge that needs to be addressed on a priority basis.

POST-TSUNAMI RELIEF AND REHABILITATION: KEY AREAS OF CONCERN

The tsunami left in its wake unprecedented destruction of life and resources and posed a huge challenge to the State and civil society, both of which have demonstrated a positive intent as well as informed and decisive action. A synergy of their mutual strengths has been an important factor in ensuring the development of a meaningful and strategic response. While response efforts in Tamil Nadu have been by and large recognized as having been comparatively effective and successful, it remains a fact that there are several issues related to the relief and rehabilitation processes that are a cause for concern, and that need to be critically delved into as an initial step towards drawing vital lessons that can be taken forward to other disaster situations.

Immediate Response and Relief

The relief phase saw the government, NGOs, local public bodies, religious groups, hospitals, etc., launch an intensive effort to rescue people, retrieve and cremate or bury bodies, clear debris, transport people to relief camps, organize medical facilities, make arrangements for food, water and sanitation, and extend psychosocial care. The Armed

Forces were brought in for search and rescue operations, and special measures were taken to avoid epidemiological crises.

While on the whole relief activities are recognized as having been effectively coordinated, there were critical gaps and problems associated with the organization and provision of relief. In the initial days, relief was given only to those registered as tsunami affected, and people who had fled their homes post-tsunami, returning after a few days, had considerable difficulty in registering themselves. Further, at the start, aid reached only fish worker communities, while several other communities who had lost their means of livelihood, but perhaps not property and life, were overlooked completely. There were reports that in many villages, even if aid was distributed equally to all, it was usually collected back later, and distributed only to those who lost assets like boats and catamarans (Sampath, 2005). Relief supplies often did not reach villages that were situated at a distance from the main road. There was also a fair amount of duplication of work and dumping of relief material due to uncoordinated relief distribution by a plethora of agencies. Needs of vulnerable groups such as women, children and the aged were typically overlooked in relief camps.

While initially schools, colleges, marriage halls, etc., functioned as relief camps, people were soon moved out to temporary shelters. With regard to temporary shelters, haste to complete construction by the 15th January, 2005, appears to have undone an otherwise well-framed policy. Typically, temporary shelters were very hot, poorly ventilated, had no flooring, inadequate water and sanitation facilities, and insufficient lighting. The shelters were built mostly in low-lying areas, and were flooded during the monsoons in 2005. A year after the disaster, many families continue to reside in these shelters, and are likely to do so for at least another three to six months. While some civil society organizations have been involved in upgrading the quality and facilities of these shelters, the overall experience has been one in which even widely known standards such as the Sphere standards have not been adhered to.

Rehabilitation–Housing Reconstruction and Livelihood Recovery

1. Housing Rehabilitation: The Relocation Conundrum

Housing reconstruction is being implemented in partnership with civil society organizations in all affected districts, except Chennai and Thiruvallur. The housing rehabilitation policy initially brought out envisaged the permanent relocation of affected communities in order to incorporate safety considerations, and in keeping with the Coastal Regulation Zone (CRZ), notification (the CRZ Notification, issued in 1991 under the Environment Protection Act, 1986, regulates the types of activities and land uses permitted in the coastal regulation zone. The notification, however, allows for fish workers settlements and associated activities in these zones).

In the face of strong opposition from civil society on the grounds that relocation would have a direct, negative impact on fishing livelihoods, a new GO was issued making relocation optional for people residing within 500 meters of the high tide line. However, people residing within 200 meters of the high tide line would not receive any government assistance for repair/reconstruction unless they relinquished the site and house, in return for a new house beyond the 200 meters zone.

The housing rehabilitation policy is evidently attempting to address several concerns, often conflicting, ranging from livelihoods and safety, to environmental and legal considerations. However, the fact that the policy excludes people living within the 200-meter zone, a section of the affected population that is extremely vulnerable given its proximity to the sea, is a matter of concern. The State would have done well, rather, to adopt a policy that allows partnership with the people and civil society organizations to create disaster resistant housing for such vulnerable populations.

2. Housing Rehabilitation: Procedural Complications, Quality and Community Participation

Acquiring land for resettlement of communities, despite a special directive permitting district administrations to acquire land through private negotiation, has been very difficult given the scarcity of land and high land prices on the coast. Difficulties in land acquisition have led to delays in allocation of habitations (for reconstruction) to NGOs, and the accompanying compulsion for people to stay for a prolonged period in the temporary shelters. To circumvent this, district administrations have been informally encouraging communities to opt for in situ rehabilitation, even within the 200 meters zone.

Quality of construction has been another issue of concern, particularly when the implementing agency hands over construction to a building contractor and removes itself from the scene, leaving room for malpractice. The government's stand that it will not provide assistance for reconstruction within 200 meters of the high tide line is especially problematic in light of quality concerns, as this implies that there are no technical guidelines, and allocation and monitoring mechanisms governing reconstruction here. When it comes to site selection, criteria regarding distance from sea and elevation are often not followed, and quality of land is at times not acceptable–for instance, wetlands and saltpan lands have been allocated in some areas. In totality, inadequate monitoring and the lack of political will to enforce minimum standards have impacted housing reconstruction negatively.

Participation of the affected community typically remains at a minimum, despite the policy recommending early allocation and linking of sites with beneficiaries to allow them to be a part of the process right from the start. Consultation processes for design, layout planning, etc., are not always followed, or involve only power centers in the community. The housing policy also favors external organization involvement over owner-driven housing rehabilitation (the rehabilitation experience after the Gujarat Earthquake of January 26, 2001, shows that owner-driven housing, with financial and technical support from, and monitoring by the State, is often the best option for rehabilitation in terms of resulting in sustainable habitations that are in keeping with people's needs). Further, the ownership of the land remains with the State Government, with the district administration reserving the right to transfer ownership to the beneficiaries as and when it deems fit.

3. Livelihoods

A range of livelihoods across the Tamil Nadu coast was affected by the tsunami. While fishery-based livelihoods were the first to be recognized as requiring rehabilitation assistance, it soon became apparent that a large percentage of people on the coast dependent on other livelihoods, particularly agriculture and informal sector occupations, were affected as well.

Despite this recognition, rehabilitation of fishery-based livelihoods (this discussion on rehabilitation of fishery-based livelihoods draws extensively from policy notes released by the NCRC, Nagapattinam) has continued to receive the maximum attention from both the State and civil society organizations. The focus in this sector has been on replacement of assets, i.e., provision of boats and fishing equipment. The combined and, at times, uncoordinated efforts of the State and civil society to supply boats to fish workers has led to an increase in the total stock of fishing boats in Tamil Nadu, with possible long-term, negative consequences for fishery in Tamil Nadu which had already been experiencing stagnation due to over-exploitation. This over-supply, often driven by the desire of civil society organizations to foster equity by giving boats to those who were earlier laborers, has also led to a shortage of fishing boat crew, with the result that young boys are being pulled out of school to work on the boats. Issues related to quality are being seen–boats are often not seaworthy, as organizations seem to compromise on quality in an effort to increase numbers. There are also issues relating to the lack of suitability of boats to local preferences and sea conditions. The fisheries rehabilitation policy also has not addressed deeper issues associated with depleting fishery resources, sea safety (such as search and rescue operations for stranded fish workers) and disaster preparedness (for instance, by means of insurance coverage).

Accurate damage estimation has been problematic[1] as a large proportion of the craft were unregistered and without any records. To circumvent difficulties created by the lack of existing data, the rehabilitation process has been largely routed through the traditional fish worker *panchayats* (these are community institutions different from the elected *panchayats)* that govern these communities. While these have greatly facilitated identification of beneficiaries and implementation of rehabilitation packages, they are not egalitarian in several respects. Women do not have a place in these traditional *panchayats*, and so are often left out during rehabilitation processes. While the *panchayat* does support them for their basic needs, this support is usually at the subsistence level.

With respect to agriculture, it is widely believed that the official estimates of affected lands are far less than the actual figures. The compensation for crop damage is inadequate, and the efficacy of the desalination packages provided by the government is still under question (civil society organizations are exploring organic techniques for restoration of agricultural lands, particularly in the Nagapattinam district). Agricultural laborers, mostly *Dalits*, have been subsumed in the larger discourse on agricultural land, with typically only landowners being considered as tsunami affected. While in some areas relief packages did reach laborers, nothing much has been done for them in terms of livelihood rehabilitation.

The situation of people engaged in petty trades, service provision, allied services such as fish curing, vending, etc., and working in saltpan lands is similar. Relief packages were announced for these groups, but little is available by means of livelihood rehabilitation, as it is believed that these dependent livelihoods would pick up as the fishing economy revived. It is also being seen that claiming compensation for loss of livelihood, particularly when one is a part of the unorganized sector, is very problematic because of issues related to evidence and proof.[2] Getting losses certified by the administration, or getting people to vouch for them during enquiries has been a difficult proposition[3] for itinerant vendors and service providers. Some civil society organizations have been involved in providing legal aid services to those who have not received their entitlements. Alternative livelihood programs are also being explored.

What cuts across the entire experience of livelihood rehabilitation is that both the government and civil society organizations have largely adopted property-centric relief and rehabilitation policies, with the result that livelihood rehabilitation packages predominantly reach only people who possess assets such as boats, land, shops, etc. Several groups that contribute to the coastal economy by providing their labor and skills have all found themselves sidelined and disregarded.

An Overview of Accountability and Transparency in Tsunami Response

The working of the various agencies involved in the relief and rehabilitation process has not been free of concerns related to transparency and accountability.

There is a widely perceived need for civil society organizations to follow rigorous procedures with regard to financial management and related transparency, particularly in the context of the large sums of money that have been received in the wake of the tsunami, a process that has been made easier by the waiver of the traditional checks otherwise placed by the Foreign Contribution Regulation Act (FCRA) of 1976. Civil society organizations have been typically reluctant to share information on resources received and utilized with other civil society actors, often saying that they are accountable to only the government.

People also voice concerns relating to issues such as poor quality and slow pace of work, non-participatory approaches and a general unwillingness to share information. The visibility[4] as well as the financial resources that tsunami related work seems to have, leads to intense competition to "acquire" villages and spend massive budgets, often in a non-consultative and uncoordinated manner (Anand, 2005). It is also troubling that despite many claims by NGOs that their focus of work and advocacy is on vulnerable groups, issues like the conditions of agricultural laborers, and lacunae in old age pension schemes have not received adequate attention in policy.

While the participation of affected people is widely upheld as a principle, in practice, however, it is not always realized due to the nature of institutional arrangements around financial reporting and obligations. One reason cited by NGOs as a hindrance in giving full play to participatory processes was the necessity to achieve targets and meet project deadlines. It also appears that lack of mutual familiarity and knowledge of local conditions makes it all the more difficult to foster mutual trust, an essential condition for effective participation. Many organizations have tried to circumvent this problem by hiring locals, ideally people from the affected community itself, as their outreach workers.

The nature of institutional arrangements governing donor-partner relationships also affects relevance (meaning the extent to which the program meets people's felt needs) and effectiveness of programs. Contractual obligations, including sanctioned budget lines and conditions regarding program type and implementation, have had significant bearing on the level of flexibility in program administration. It is commonly seen that there is minimal scope to make alterations in the program, or make mid-course corrections because of various conditions imposed by donor agencies.

Relevance of programs is also affected by the enthusiasm of many organizations to apply new knowledge without researching it carefully, and to bring about social change in the community.[5] Further, one finds that many of the agencies engaged in post-tsunami relief and rehabilitation do not have the experience, expertise or knowledge of the coastal economy, social systems and structures required to design and implement meaningful programs. When such organizations "influence" their local partners to take up "specific" areas of work or modes of implementation, intervention often becomes inappropriate to local conditions.

What emerges from these experiences is the value of different types of civil society organizations confining themselves to clearly defined roles. There seems considerable merit in the argument that donor agencies that are not familiar with local conditions should not take on the role of a direct implementing agency. Ideally, while a development support agency brings in the financial resources, a local development organization must work to facilitate the rehabilitation process in partnership with the community.

With respect to the State, the administration seems to have earned a fairly clean chit in terms of accountability and transparency in its working. While this could be attributed to the fact that the government has, for the most part, stayed away from direct, beneficiary oriented implementation of rehabilitation, it is also true that several mechanisms have been set up to ensure a certain level of transparency in the disaster response. For one, the government has taken a deliberate decision to partner with civil society organizations in the rehabilitation process, and this, along with several other monitoring measures, has gone a long way in minimizing corruption in the delivery of aid.

With respect to financial transparency, the government has exhibited a great deal of willingness to share information. However, some ambiguity exists in civil society regarding the exact quantum and utilization of funds received from various sources such as the World Bank, Asian Development Bank, and the Central Government. The government, being obliged to set the highest standards of transparency and accountability, should take immediate steps to clear this ambiguity, by perhaps issuing a White Paper on this question.

VULNERABILITIES AND EXCLUSION

The ongoing process of relief and rehabilitation has seen the exclusion of several groups of people because of a complex matrix of reasons relating to factors as diverse as inherent vulnerabilities and active discrimination, as well as systemic deficiencies.

The special needs of vulnerable groups such as women, children, the disabled, and the aged were not considered in the relief phase. For instance, the relief packages distributed did not contain even children's clothing. Older people found it difficult to collect relief packages in camps as they were pushed around. A study conducted by HelpAge International during the relief phase reveals that a main reason underlying the invisibility of older people is the absence of data disaggregated by age (HelpAge International, 2005). Women faced several problems, including those related to lack of privacy and appropriate sanitation facilities.

The Tamil Nadu government has, however, done well to respond to civil society's concerns about child trafficking in the guise of adoption after disasters, and has directed that all affected children are to remain in their home districts, and that all orphans without caretakers are to be housed only in government orphanages.

With regard to the elderly, the government's old age pension scheme, that was extended with immediate effect to all eligible tsunami affected people who were not already covered, is severely lacking. Under this scheme only people above the age of 60 who do not have a son above the age of 18 years are eligible for pension. The scheme completely ignores the fact that many elderly, destitute people are estranged from their families and sons, and so are as helpless and vulnerable as those without sons. Further, the fact that the policy mentions sons rather than daughters is contrary to the principle of gender justice that should be the cornerstone of an equitable rehabilitation process. The disabled receive no mention in the rehabilitation policy, other than directions issued ordering the distribution of aid devices.

There are reports of *Dalits* facing discrimination during the relief phase, and being denied even food aid on the grounds that there were no deaths among them (Alternative Law Forum, 2005). In the rehabilitation phase, the prospect of relocation is leading to caste related conflicts. With certain fish worker communities opting for relocation, identifying suitable sites becomes problematic when *Dalit* communities are found between the new site and the sea, as traditionally *Dalit* settlements are always leeward of fishing villages.

Women's work has not been recognized by rehabilitation programs (Oxfam, 2005). Livelihood restoration measures have focused on replacement of assets, thus excluding women who rarely possess assets in their names. Often the disbursement of relief and rehabilitation packages is controlled by fish worker traditional *panchayats* that are dominated by men, leading to further exclusion of women. Even when women do get compensation and relief packages, other family members typically take over the resources that come in (Manecksha, 2005). It must be noted here, however, that the government has taken several measures to incorporate women's concerns in the rehabilitation policies–for instance, all houses are to be registered in the names of both the husband and wife; any transfer of the wife's share to the husband is void.

The case of the *Irulas*, a semi-nomadic *adivasi* (tribal) community, reveals with a great deal of clarity an important factor influencing exclusion in disaster response. The *Irulas* (several members of which are engaged in backwater/catamaran fishing near the shore, or work as casual labor on fishing boats) as a group were, and continue to be, overlooked in the relief and rehabilitation processes.

As a community, the *Irulas* have long been invisible given their nomadic habits, and scattered, sparsely populated settlements. They rarely possess proofs of identity such as ration cards (an official document issued by the Government that entitles holders to subsidized food from the Public Distribution System), voter's identification cards, or caste/community identity certificates (entitling members of marginalized groups to benefits of affirmative action). Many of the households do not even appear in the Census as the *Irula* settlements are usually located far from the main roads, and people are usually not available at home during the day.

After the tsunami, *Irulas* whose houses were affected had nothing to show–they lived in huts that were completely washed away, leaving no signs of the destruction that had been wreaked. Their settlements were dispersed, so identification of the affected and distribution of relief became huge tasks. A variety of evidence was required to get compensation–a house on *patta* (a title deed) land, recognition of the village by a *panchayat* (*Irulas* live in dispersed, isolated clusters, unrecognized by the VAO or a *panchayat*) and other government identification cards–none of which was available with them. The *Irulas* are also, by nature, a reserved community, and are reluctant to even speak about the losses they have suffered in the tsunami to outsiders. In short, they lack the ability to sell themselves, and this when contrasted with the highly organized and articulate fish worker community, puts the *Irulas* even more at a disadvantage. It has also been noticed that even civil society organizations are not very eager to support *Irula* rehabilitation possibly because their settlements are scattered and sparsely populated. It is easier to get numbers (in terms of families and people), identify beneficiaries, distribute relief and rehabilitation packages, and raise resources in the densely packed fish worker settlements.

Understanding Exclusion: The Politics of Information and Definition

The Tamil Nadu experience shows that exclusion in post-disaster situations is linked to an interaction of factors such as inherent vulnerabilities, data integrity, and definition of "affected."

In the first days after the tsunami, no relief reached a particular Hindu village in the predominantly Christian coastline of Kanyakumari, as the district administration depended on the Church for information about affected settlements (the Catholic Church is a very well-established institution in this area, and played a crucial role in coordinating relief and rehabilitation processes). This village did not figure in Church records as the Church maintained detailed records only of the Christian settlements. As the administration did not possess readily accessible, accurate geo-demographic data, it relied, and continues to rely, on the Church for records, thereby creating a situation in which there are bound to be exclusions. The neglect of the Irula community can also be attributed to similar reasons–only a hazy, distorted picture of these people exists with the government, and hence its inability to proactively identify them as "affected" and address their concerns.

As can be seen from these instances, and from the situation on the ground, the underlying issue emerges to be one of data integrity. Detailed and accurate demographic (inclusive of socioeconomic and geographic information) data and other relevant records are not available with the government, and where available, are often not accurate.

This deficit in data integrity is exacerbated by a skewed understanding of the term "affected" that is based on the premise that only settlements that were damaged by water, and that only people who had lost lives and property, were "affected." This visibility-influenced definition has in turn directed the manner in which compensation, relief and rehabilitation processes were and continue to be designed and implemented–there is stress on provision of evidence that one has lost assets. Added to this is the flawed conceptualization of the "coastal community" as a homogeneous whole (namely fish workers), rather than a vibrant system of inherently different groups, held together by geo-economic links.

While corrective measures have been taken to widen the scope of the definition of "affected," they have not been completely effective in reaching excluded groups, as most systems and mechanisms governing the relief and rehabilitation process are still based on the original understanding of affected, as well as on the initial impact assessments and registrations of affected people.

TSUNAMI RESPONSE IN TAMIL NADU: CONSOLIDATING THE EXPERIENCES

It is clear from the Tamil Nadu experience that a responsive and proactive government, that devolves powers and responsibilities to the administration, is a necessary condition for effective disaster response. Further, institutional autonomy, and decision-making free of political interference, are crucial in fostering people-sensitive and effective relief and rehabilitation.

It also appears that a lot depends on the perspective, approach and capacity of individual officers in key positions; however, this overreliance on the ability of the individual occupying a particular office is not without its dangers. It is necessary to examine means of extrapolating the learnings of such officers to an institutional level in order to create a common skills and knowledge base.

While an efficient and non-corrupt administration is a necessary condition for effective disaster response, particularly in the relief phase, a much deeper engagement is needed with local elected institutions–the elected *panchayats*–to ensure equitable and inclusive rehabilitation, and to counter several of the problems seen commonly. The 73rd Constitutional Amendment vests with the village *panchayat* the obligatory function of disaster management and preparedness; *panchayats* possess the potential to implement the rehabilitation process at the grassroots level, and are well-placed to foster community participation. They can also play a key role in creating detailed demographic databases, and are perhaps the institutions best suited to monitor reconstruction, as well as incorporate long-term development goals in the rehabilitation agenda.

However, in practice, one finds that in most post-disaster situations, as in the case of the tsunami, elected *panchayats* (as different from traditional *panchayats*) are not given the space and resources to facilitate the reconstruction process. Despite these obstacles, in Tamil Nadu there is newly emerging evidence (Parasuraman, 2006) that in places where *panchayats* have been active, the recovery process has progressed much faster and in a far more accountable manner. It is therefore vital that the involvement of these local bodies in disaster response is institutionalized in practice, and that they are equipped with the necessary capabilities and resources.

This study of tsunami response in Tamil Nadu also shows that a vibrant interface and partnership between the State and civil society is critical to ensure that people's concerns are addressed by both policy and practice, as well as to minimize corruption. This synergy needs to be fostered by both the State and civil society proactively. Also important is the need for effective communication channels, coordination, dialogue and information among and within civil society organizations to avoid duplication, wastage of resources and competition, and to optimize individual efforts.

With regard to the several instances of poor quality of intervention, it is inexcusable that such issues continue to plague disaster response despite the widespread knowledge and awareness that exists about standards in humanitarian aid, for instance, the Sphere Standards. What is patent, however, is that, given that post-disaster situations are characterized by flux, a great deal of flexibility in terms of rehabilitation policy, the nature of interventions, budgets and timelines, and even program objectives, is required to allow for the adapting of response interventions to address ground level realities as well as the true needs of people.

Further, there is a strong case for civil society organizations to put in a lot more effort in fostering transparency in their interventions. At the end of the day, all funding they receive and all programs they administer are in real terms meant for the people. Therefore, they should be first in line to share with others all details of their functioning, and in fact should welcome the involvement of the affected community in monitoring their interventions and programs.

Moving on to the questions of exclusion, what counts is the capacity of the system to deal with this exclusion, and for this, accurate and reliable disaggregated data on various lines is crucial, as is the willingness and capacity to alter policy to suit changing needs. Post-tsunami, it is strikingly clear that there is a pressing need for detailed and inclusive geo-demographic and socioeconomic data collection, and vulnerability mapping as a resource base for disaster preparedness. The importance of avoiding rigid and unalterable definitions of "affected" is also evidenced by this experience.

The importance of disaster preparedness has never been so clearly brought out as in the case of the tsunami. Preparedness is needed at various levels; for one, there needs to be preparedness to respond, in terms of institutional structures and systems, personnel as well as material resources. Community preparedness is another crucial aspect. The first few hours after the tsunami showed that it was the community itself that was the first to respond, as in the case of all disasters. Communities need to be prepared so that not only are the impacts of disasters mitigated, but also so that they are knowledgeable about how to respond. Preparedness also includes addressing vulnerabilities: development of capabilities, reducing poverty, access to insurance, early warning systems, etc., all form a part of this. And at the broadest level, the need for a holistic and integrated policy on disasters that covers all aspects of prevention, preparedness, mitigation and response cannot be ignored.

NOTES

1. The process of delivering relief and rehabilitation in a manner in which resources are spent judiciously, while ensuring that all affected people get their entitlements has posed quite a challenge to the government. Given the absence of reliable population data and records, the primary difficulty lay in identifying beneficiaries and verifying whether their losses were genuine. The government has relied heavily on community-based institutions to get data on damages and affected people.

2. In the agricultural sector, compensation for loss of livestock is sanctioned only if the claimant has photographic evidence of the death of the animal–in the wake of the tsunami, most bodies were washed away, and those that remained were given a hasty burial to prevent disease outbreaks, making it next to impossible to produce such proof.

3. A case in point is the plight of *Narikurava* (gypsy) vendors in the tourist town of Mahabalipuram near Chennai city. The tsunami waters washed away their wares–artifacts and costume jewellery. But they have been unable to access compensation as

there is no one to vouch for their losses–not only do they keep moving, they are also viewed with distrust and contempt by the mainstream.

4. It was pointed out by many that NGOs prefer to go to Nagapattinam, Cuddalore and Kanyakumari as they receive media attention. In contrast, in areas like Ramanathapuram or Vizhupuram, Thiruvallur, and Kanchipuram, one finds very few NGOs, and also only local NGOs that do not have much capacity to raise or spend money.

5. A case in point is the virtual flood of surveys and administration of psychosocial interventions that followed the tsunami. While many of the psychosocial care interventions were sensitively designed, there were others that were not so. Several months later, affected people still recount the trauma of having to repeat the same stories, draw the same pictures and listen to strangers asking one to "cry and get it out," in many cases, several times over with different agencies and organisations. Another instance is that of the oversupply of boats caused by the desire to foster equity that in turn has led to young boys being removed from school to work as crew.

REFERENCES

Alternative Law Forum (2005). *Towards ensuring food security for the Tsunami affected: A position paper.* Retrieved October 30, 2005 from http://www.altlawforum.org/Resources/Tsunami/Positionpaper.doc

Anand, S. (2005). *Everybody loves a good Tsunami.* Retrieved December 12, 2005 from http://www.outlookindia.com/full.asp?fodname=20051121&fname=Tsunami+Relief+%28F%29&sid=1

Government of India (2005). *Special situation report–35.* Retrieved December 03, 2005 from http://www.ndmindia.nic.in/Tsunami2004/sitrep35.htm

HelpAge International (2005). *The impact of the Indian Ocean Tsunami on older people: Issues and recommendations.* Retrieved November 30, 2005 from http://www.helpage.org/Resources/Researchreports#swHC

Manecksha, F. (2005). *Restructuring society post-Tsunami.* Retrieved October 29, 2005 from http://www.indiatogether.org/2005/jun/wom-aftertsu.htm

Oxfam (2005). *The Tsunami's impact on women.* Retrieved October 29, 2005 from http://www.oxfam.org.uk/what_we_do/issues/conflict_disasters/downloads/bn_tsunami_women.pdf

Parasuraman, S. (2006). Personal communication on January 4th, 2006.

Sampath, B. (2005). *Tsunami relief work report: A perspective report.* Retrieved October 12, 2005 from http://www.pratham.org/news/tsunami.doc

Srinivasan, K., Nagaraj, V. K., & Venkatesh, K. (2005). *The state and civil society in disaster response: An analysis of the Tamil Nadu Tsunami experience.* Mumbai, India: Tata Institute of Social Sciences.

Tsunami–The killer waves: Damages (2005). Retrieved December 03, 2005 from http://www.tn.gov.in/tsunami/damages.htm

doi:10.1300/J198v05n03_04

After the Wave–
The Centrelink Social Work
Response Offshore

Caroline Manning
Stewart Millar
Tania Newton
Sharon Webb

SUMMARY. Communities worldwide were affected by the tsunamis that struck a dozen Indian Ocean nations on 26 December 2004. More than 300,000 people were killed or left missing as a result of the tragedy. The Australian Government played a key role in the disaster response overseas and in Australia. As part of a whole-of-government response to the crisis, eight Centrelink Social Workers were deployed to tsunami-affected areas for two weeks after the initial disaster to provide personal

---

Caroline Manning, BSW(Hons), is Business Manager, Social Work Services, Centrelink, P. O. Box 7788, Canberra Business Centre ACT 2610, Australia.

Stewart Millar, BA, is Senior Social Worker, Launceston Call Centre, Centrelink, P. O. Box 586, Launceston Tas 7249, Australia.

Tania Newton, BSW, is Business Manager, Social Work, Area North Australia, Centrelink, LMB 86 Darwin, NT 0811, Australia.

Sharon Webb, BSW, is Business Manager, Social Work, Area South Metro, Centrelink, Locked Bag 7004, Liverpool BC NSW 1871, Australia.

Special thanks go to Martin Gribbon, Desley Hargreaves, Heather Peasley, and Charlie Wickins for their comments and contributions to this article.

[Haworth co-indexing entry note]: "After the Wave–The Centrelink Social Work Response Offshore." Manning, Caroline et al. Co-published simultaneously in *Journal of Social Work in Disability & Rehabilitation* (The Haworth Press, Inc.) Vol. 5, No. 3/4, 2006, pp. 81-95; and: *Asian Tsunami and Social Work Practice: Recovery and Rebuilding* (ed: Ngoh Tiong Tan, Allison Rowlands, and Francis K. O. Yuen) The Haworth Press, Inc., 2006, pp. 81-95. Single or multiple copies of this article are available for a fee from The Haworth Document Delivery Service [1-800-HAWORTH, 9:00 a.m. - 5:00 p.m. (EST). E-mail address: docdelivery@haworthpress.com].

Available online at http://jswdr.haworthpress.com

doi:10.1300/J198v05n03_05

support to Australian victims and their families. This paper presents the service delivery framework adopted offshore and delineates the role undertaken by the social workers. Drawing on their practice experience, the paper examines some of the challenges, dilemmas, learnings and implications for future service delivery. doi:10.1300/J198v05n03_05 *[Article copies available for a fee from The Haworth Document Delivery Service: 1-800-HAWORTH. E-mail address: <docdelivery@haworthpress.com> Website: <http://www.HaworthPress.com>*

KEYWORDS. Disaster recovery, whole of government, tsunami, personal support, natural disaster, disaster response

INTRODUCTION

Within hours of the tsunami striking on 26 December 2004, the Australian Government mounted a whole-of-government emergency response to the crisis. Similar to the 2002 Bali bombings, the tsunami was a large-scale disaster that affected the whole nation and required both a domestic (i.e., Australia based) and international response. As lead agency, the Department of Foreign Affairs and Trade (DFAT) coordinated the offshore deployment of a number of Commonwealth agency representatives including the Australian Federal Police (AFP), Emergency Management Australia (EMA), the Australian Defence Force (ADF) and Centrelink. This coordinated and integrated approach built on and was informed by the key learnings from the whole-of-government response to the 2002 Bali bombings (Management Advisory Committee, 2004). The purpose of this article is to document the social work practice, interventions, dilemmas and learnings of the offshore experience. Some attention will be given to the response provided after the 2002 Bali bombings as it informed, influenced and shaped the tsunami disaster recovery experience (Gribbon, McLennan, Millar, & Newton, 2004).

Background

As one of the Australian Government's main human service delivery agencies, Centrelink has a long history in disaster response and recovery in Australia. Social Workers have been integral to this response

(Centrelink, 2005a, 2004, 2003, 2002, 2001, 2000, 1999). Following the 2002 Bali bombings and at the request of the Bali Commonwealth Inter-Agency Taskforce, the Centrelink Social Work Service established a Family Liaison Officer (FLO) Program to provide an ongoing liaison and support service to affected families and survivors. This was supplemented by Social Work assistance provided via the Bali Hotline and "Meet and Greet" arrangements at the airports. Additionally, a direct and personalised client service for victims and family members was provided offshore during the trials of the accused and at commemorative ceremonies (Gribbon et al., 2004). This offshore service delivery was unprecedented for the Social Work Service.

SERVICE DELIVERY AND PRACTICE FRAMEWORK

Following the tsunami, Social Work services were once again provided in Australia to victims and family members via a 24-hour/seven-day-a-week Tsunami Assistance Hotline and the FLO Program. People evacuated and returning from the affected areas were met by Centrelink staff at airports to provide referrals, support and advice. Linkages were made with State and Local governments and community-based support agencies. Additionally, eight Social Workers were deployed to tsunami-affected areas in Thailand and Sri Lanka to undertake the role of Family Contact Officers (FCO). This comprehensive and coordinated framework ensured the seamless availability and provision of personal support and assistance to victims and family members immediately following the disaster and beyond; and thus, complied with the principles of disaster recovery management (Emergency Management Australia, 2004).

Gribbon et al. (2004) highlighted several factors that deemed the service delivery unique following the 2002 Bali bombings. These included:

> . . . the nature and scale of the tragedy; the deliberate terrorist attack on innocent, mostly young holiday makers; the seven month gap between the bombings and our "on site" intervention; the high profile media coverage; the community outcry . . . the potential for future attacks and the offshore location. (p. 1)

Clearly, the tsunami presented a different context: the sheer scale, size, geographical spread and number of casualties; the "natural" aspect of the disaster; the immediate emergency response deployment; the deployment to several locations; and the relatively protracted "tour of duty." Notwithstanding these contextual differences, the established service delivery model provided clear practice imperatives and a fundamental framework for the offshore emergency response.

As mentioned earlier, the response involved several distinct agencies that provided particular services and expertise under the leadership of DFAT. As such, an interdisciplinary approach within and across organisational boundaries and protocols was required. Constructive working relationships formed quickly were essential. The Centrelink Social Workers were deployed in teams of two, with gender balance largely maintained, to Phuket and Bangkok, Thailand, and Colombo, Sri Lanka initially. Personnel were subsequently deployed to Krabi, Thailand also to respond to emerging needs.

FAMILY CONTACT OFFICER ROLE

"Recovery from disaster is an enabling and supportive process which allows individuals, families and communities to attain a proper level of functioning through the provision of information, specialist services and resources" (Emergency Management Australia, 2004, p. 105). The primary role of the Family Contact Officer was to provide personal support and assistance to Australian survivors and family members and to " . . . facilitate whatever has to happen for them in as seamless a way as possible" (Centrelink, 2005b, p. 6). The interventionist roles encompassed but were not limited to those delineated in foundational Social Work texts (Howe, 2003; Compton & Galaway, 1999; Garvin & Seabury, 1997).

The disaster context necessitated a flexible and responsive assumption of roles depending on client needs at different points in time. Further, there was a need to establish structures in response to emerging needs. All interventions were driven by client identified priorities that ranged from the fundamental and practical to the spiritual. The coping strengths, capabilities, resources and natural support systems of those affected were identified and harnessed (Mutton, 2005; Cowger, 1994). Throughout, it was important to encourage the articulation and ventilation of feelings; bear witness; assist in identifying needs and concerns; provide support and ensure client self-care.

Specific aspects of the role included:

- Contact with families on arrival and departure from Phuket and Bangkok airports. The familiarity of the Australian flag and personnel clearly provided many affected people with a sense of comfort and support in an otherwise disorienting environment. Contact at the airports ensured that clients received support at key "stopovers" on their journeys. It was important that clients received a seamless service. To this end, information about the location of families was passed to relevant personnel and details of the personnel who would meet them at each stage of their journey home were provided to families and victims.
- Support to families searching for the missing and identifying deceased loved ones.
- Providing information to families pertaining to the scale and nature of the disaster as well as information about normal trauma reactions and the exploration of management strategies. Preparation of the families for their likely experience, including potential physical and emotional reactions, given the confronting nature of the disaster, was important. This was particularly so when families were preparing to search the temporary morgues and photos of deceased victims for their loved ones.
- Facilitating formal and informal linkages for clients. In the post-disaster context, this involved identifying immediate and available support options such as forensic, consular, and pastoral care services. A key strategy in developing informal linkages was the facilitation of support networks amongst affected people (Dyregrov, 2001). For example, information sessions and opportunities for informal gatherings were offered. In addition, it was important to give people a sense of continuity and connection with support services on return to Australia. This included other Centrelink services and community supports.
- Accompanying families to various sites that they wished to visit and offering support where appropriate. Locations included crisis command centres, hospitals, temporary morgues, and public memorial services.
- Assistance to families in the preparation and planning of private family memorials.
- Assistance to family members with coping and management strategies in relation to preexisting family relationship issues; entrenched family secrets and blame; and, competing claims to belongings and remains that were exacerbated by the crisis. Knowledge of family dynamics and intervention techniques, skills

in crisis intervention, and the knowledge of disaster effects enabled the tailoring of interventions in these situations (Dyregrov, 2001).

- Assessment of the need for more specialist/intensive intervention due to preexisting mental health risks and conditions to ensure that client long-term well-being wasn't compromised.
- Advocacy and problem-solving in relation to flight schedules, flight upgrades for people with specific needs, facilitating timely repatriation of deceased loved ones; and, moderating challenging and sometimes unfeasible expectations of other agencies and clients.
- Mediation and conflict resolution. Typical sites of conflict were intra-familial; between affected people and government agencies regarding the negotiation of policies and procedures; the media; and volunteers. A particular example is the complex and time consuming nature of the victim identification process which exacerbated the distress of some family members.
- Supportive counselling of families in relation to the loss of loved ones and property.
- Assisting families to manage potential encounters with the media. Media interest was excessive and at times invasive.
- Establishment and staffing of a 24-hour/seven-day-a-week FCO crisis line.
- In all disaster situations, practical assistance is a key component of effective recovery. If you can't do anything else, attend to the practical. Specific examples from the tsunami response included: locating hospitalised survivors; facilitating contact with family members at home; providing information and advice on local infrastructure such as mobile phones, transport, accommodation and resources such as flowers; language and translated information; cultural practices and sensitivities; taking of, and provision of photographic records for families; meeting administrative requirements such as addressing Centrelink payment issues; and claims for funeral assistance. Victims and family members can experience an overwhelming sense of guilt associated with attending to practical issues in the midst of tragedy. It was important to encourage the maintenance and reestablishment of daily routines.

In establishing the FCO role, articulation and promotion of the service offer was essential. As a result, the Social Work skills, knowledge and resources were sought to address some of the otherwise unmet needs. While these requests were secondary to the core role, the FCOs responded if and when resources permitted. These requests included:

- Support to staff; volunteers and expatriates. Given the magnitude of the disaster, many of these people experienced the tragedy within their own families and communities, yet they continued to assist with the recovery. At times it was the nature of recovery tasks that necessitated the need for support, e.g., removal of large numbers of bodies from the sea.
- Consultancy to Embassy staff and other agencies. This was not an obvious or predetermined role but one that emerged due to the nature and scale of the disaster, and in some areas, the lack of other support services. Advice was sought on strategies to address issues associated with impacted staff, and assistance with development of school-based programs for affected students.

KNOWLEDGE AND THEORY

A range of theories and knowledge informed the interventions of the FCOs including crisis intervention, brief casework, empowerment, strengths-based, grief and loss, narrative, cognitive-behavioural, trauma responses, family and groupwork, systems and disaster recovery (Payne, 2005; Cowger, 1994; Austin, 1992; Hodgkinson & Stewart, 1991). Particularly, crisis intervention provided the conceptual framework for the brief, focused interventions of the tsunami recovery with workers tailoring their interventions to individual clients, situations and particular problems. While the scope of this paper does not permit further elaboration of the theoretical framework, the importance of it is not to be underestimated.

OCCUPATIONAL HEALTH AND SAFETY

Occupational Health and Safety issues were presented for staff both prior to departure and while offshore, examples of which are outlined below. Readily accessible professional supervision was critical in ensuring that strategies were effected to reduce the occupational and safety risks, and that staff health and well-being were not compromised.

Due to the limited time available between notification of deployment and departure times, the closure of most medical services during the holiday period, shortages of certain vaccinations and the receipt of conflicting medical advice, staff received different vaccinations and in some cases, none at all. Access to vaccinations offshore was limited also and as a result, courses of treatment were not completed.

Staff members were mindful of the risk of falling ill and the challenges associated with obtaining access to appropriate medical information and treatment in disaster areas where medical services were already stretched. In Thailand, particularly, the potential impact of exposure to decomposing bodies and the threat of disease was considered. The geographical dispersal of personnel in small teams also meant that replacements could take several days to arrive. As a result, commitment to each other and not wanting to let colleagues down pervaded the team.

Access to safe, reliable and responsive transport was limited, particularly in the early response phase. Due to the magnitude of the disaster, the local infrastructure was largely exhausted and as a result compromises were made. For example, in Thailand, workers elected to travel in the tray-back of a utility truck in order to reach various locations.

Communications infrastructure was also compromised post-tsunami. Mobile phone coverage was not comprehensive and the network was often heavily congested. On occasions when calls were connected (to the intended recipient) they invariably "cut out" prematurely. Besides the obvious frustration, the potential existed for staff to be left feeling isolated from other staff and services. It is for this reason that staff largely worked in pairs.

Typically, staff worked extremely long hours throughout the recovery. An "on call" system was established which meant that they were available to victims and their families 24 hours each day. Time zone differences also meant that calls were often received from Australia outside of regular hours. Consequently, planned "downtime" was a challenge.

SELF-CARE AND SUPERVISION

Given the abovementioned "stressors" and the disaster recovery environment, self-care strategies and professional supervision were essential (Canfield, 2005). Like all people affected by disasters, workers used a variety of techniques to deal with the experience. Ensuring adequate rest was important to compensate for the long hours and maintain energy levels, along with strategies such as exercise, humour and other practices.

Among the most important, was the use of supportive personal and professional relationships. Daily calls to home in addition to daily briefings with the team leader and professional supervisor were critical elements in worker well-being during this disaster. An informal "end of day" debrief and review that facilitated reflective practice was undertaken in each location. Further, staff received an opportunity to debrief with the Team Leader before heading home and participate in a whole team debrief within four weeks of all team members returning home; and access to an Employee Assistance Program.

LOGISTICAL AND PRACTICE CHALLENGES

Given that this was the first time Centrelink was asked to provide personnel to contribute to the whole-of-government response offshore immediately following the disaster, there was a challenge in identifying suitable staff members who were available at very short notice (e.g., four hours). This was complicated by the tsunami occurring during the Christmas holiday break.

There were numerous country-specific and cross-cultural issues that needed to be worked through in Thailand and Sri Lanka. These included visa requirements, language, access to interpreters, transport and communications infrastructure, currency, traditional burials, and other local knowledge, customs and practices.

The positioning of the FCO Service within the interdepartmental response and across three locations in Thailand (Bangkok, Phuket & Krabi) was an inherent challenge. The structure of consular services influenced how workers responded in each location. For example, services in Bangkok were largely provided from within the Embassy, at Bangkok International and Domestic airport terminals, and hospitals. Services in the disaster locations of Phuket and Krabi were provided at the consular crisis centres set up in hotels; community crisis centres; local hospitals and temporary morgues; client accommodation; and public meeting places. Further, the FCOs in Phuket were initially linked to the medical team and this influenced the early Social Work response. This was largely due to the early deployment of FCOs to Phuket. As the response and recovery progressed and medical personnel departed, renegotiation and positioning of the FCO role occurred. Similarly, the earlier allocation of roles to Embassy staff and other agencies in Bangkok prior to the arrival of the FCOs influenced their service response. Within this, though, the service offer and core role of the FCOs remained consistent throughout. Services in Sri Lanka were largely provided adjacent to the Australian High Commission, in coffee shops and people's homes.

As mentioned previously, the location of services across three locations in Thailand provided logistical challenges. Families typically moved between Bangkok, Phuket and Krabi when searching for loved ones. The presence of personnel in each location ensured personal support to clients throughout, though it meant that several workers were involved with individual families. To this end, smooth "handovers" were critical to ensuring a seamless service.

Language and nomenclature emerged as a practice challenge that required quick resolution. Use of the terms "social worker" and "counsellor" inadequately defined the role and potentially alienated clients. Both terms meant different things to different people (other agency staff and clients alike) and in some cases assumed a lack of coping on behalf of clients. Implicated were notions of being "miracle workers" and "making things better" and as such both staff and clients risked being set up to achieve unrealistic outcomes. The title of Family Contact Officers was adopted in order to facilitate client responsiveness, accommodate the flexibility of the role, and provide clarity. Within this, though, it was critical that the FCOs articulated and demonstrated their practice skills and knowledge when sought by other agency staff and clients.

ETHICAL DILEMMAS

In the interests of developing new knowledge, the identification and documentation of ethical dilemmas and, where possible, resolution, are important. They are expanded on below:

- Similar to Gribbon et al.'s (2004) experience in Bali, the media were enthusiastic to access stories about the role of the Family Contact Officers and their work with affected families. Confronting the FCOs was the challenge of finding the balance between offering insights to the general public, respecting the privacy of those impacted, promoting the service, and minimising the vicarious trauma for those exposed to the subsequent media coverage. Media requests often posed an imposition on the role and functions to be fulfilled. In order to minimise vicarious trauma, focus was given to the role and the nature of the work rather than to sensationalism and stories of horror. Emphasis was also given to the positive aspects of the disaster such as the generosity, collaboration and communality of people.

- There was an emerging tension between offering services assertively and waiting for potential clients to initiate contact. It was concluded that this was a false dichotomy as there was always an opportunity for responsiveness to people's practical needs which allowed for relationship building and potential engagement. This was consistent with the multifaceted aspects of the role as referred to earlier.
- There were inherent challenges in providing a coordinated, seamless and holistic whole-of-government response to Australian victims and families offshore, some of which have been outlined earlier. While under the direction of DFAT, organisational boundaries and protocols naturally informed and influenced various agency workers. At times this presented a dilemma for workers–to what extent should information be communicated to the parent agency? The establishment of clear reporting and accountability lines was critical. Further, clarity and delineation of roles was important given the presence of various agency staff members who had similar capacities including counselling, support, investigative liaison and pastoral care.
- FCOs were always conscious of being part of a well-resourced service entering countries and cultures within which there were potentially limited and exhausted services. While the central mandate was to offer services to Australians, the opportunity to assist some locally engaged staff of the Embassies arose. The response required awareness and sensitivity to these issues.
- In view of needing to establish rapport quickly and given the time-limited nature of interventions, flexibility in approach and venue choice was necessary. This often meant that interventions took place with families in hotels, coffee shops and other informal spaces. Given the uniqueness of the situation, the parameters within which services are regularly offered did not usually apply. This required attention to issues of confidentiality, potential intrusiveness of the media and "disaster tourists."
- Complex family dynamics are exacerbated in times of trauma and this sometimes led to dilemmas around meeting the competing needs of family members (e.g., which family member has the right to decisions regarding deceased loved ones; attendance at private memorials).
- It was important to balance availability and accessibility to other agency personnel without "getting in the way." The disaster re-

sponse fully utilised the DFAT resources and it was important that additional unnecessary demands were not placed on staff. Again, positioning was important to ensure that the FCOs were linked in to be effective.

- In Sri Lanka it was suggested that FCOs should attend the devastated areas to get a sense of the experience of people exposed. This raised issues of spending large amounts of time off-line while travelling long distances for what were considered to be questionable outcomes. Further, there was a sense in which this was considered to be voyeuristic and would probably not add to the service delivery.
- Disasters by nature require flexibility and responsiveness to needs and demands as they arise. Critical awareness of role boundaries and self-care is essential. The need to establish credibility should not come at the expense of upholding social work ethics, values and practice standards.
- Inherent in the whole-of-government approach was the need to share information across agencies. However, this raised issues in relation to client confidentiality, information recording, and the release of information.
- At times there were tensions for workers between needing to meet administrative and reporting demands and responding to clients' needs. Problem-solving, negotiation and flexibility were key skills and attributes to resolving these dilemmas.

LEARNINGS AND BEST PRACTICE

It is through learning from previous crises that preparation and planning for the future occur (Management Advisory Committee, 2004). One of the key learnings from the tsunami recovery experience was the need to have a team ready to immediately respond and the availability of a back-up team on standby in Australia. Given the nature of the work and the breadth of the role, there is a need to identify, prepare and equip suitable staff. In addition to professional knowledge and skills, desirable qualities of disaster response workers include:

- The ability to deal with complex problems and client circumstances in a confused and rapidly changing environment;
- A high degree of personal resilience and energy;
- The ability and capacity to engender confidence among staff and clients;
- Political insight, intuitiveness and astuteness;

- Sensitivity to people and work demands and a strong belief in the integrity and rights of individuals;
- Adaptability and flexibility;
- Resourceful and an entrepreneurial approach to problem-solving and results (Emergency Management Australia, 2004).

Team members' vaccinations and passports need to be up-to-date and in the latter case, also satisfy visa requirements. Deployment schedules need to be prepared prior to departure to ensure that team sizes are not reduced below two people and that back-up staff are readily available. Deployment schedules need to be flexible to accommodate varying scales of disasters and deployment destinations. Team size has a direct impact on being able to schedule "downtime." Larger teams allow for rotation of "on call" duties and access to planned time off. This is significant in maintaining energy levels and reducing the possible risk of burnout. Teams should ideally consist of male and female combinations to reflect the client base. Wherever possible, worker positions should be backfilled at home to ensure that the workers can focus fully on the job at hand.

In the context of travelling to a new country and the uncertainty of how things might evolve, travelling with or being met at the airport (by a coworker or another agency representative) was reassuring and eased some of the initial uncertainty.

In the initial stage following a disaster, it is important to be mindful of the frequency of staff rotation. The tsunami response indicated that three weeks is the maximum period that workers should continue in the front line. This, however, needs to be balanced with the need for continuity for families and the overall duration of the service response.

Ready access to key information and resources in an emergency response is critical: for example, consistent background reading for team members, business cards, information kits for our clients, pamphlet proformas, brief role statements, first aid kits, stationary and blank identification cards. Obviously the content of the kits will change with different responses; however, experience was that these kits should be available on an ongoing basis.

The instability of communications infrastructure and cost implications highlighted the need to know how to use text messaging. Prior to departure, practical considerations also need to include accessing relevant information such as geographical maps, international phone codes, global roaming facilities and mobile phone chargers.

Regular debriefing, reflective practice and documentation of learnings and experiences are key elements of effective disaster response. Often in crisis situations, documentation in particular is one thing that assumes a lower priority because of competing demands. Notwithstanding, it is a practice imperative that this maintains a high priority.

Another learning was of the need to be adaptable and flexible in relationships with others. During the tsunami recovery, informal relationships with clients were found to be more appropriate and effective than a traditional/formal counselling role. "One thing that practitioners have learnt over the past couple of decades is that focusing on the psychological needs of individuals too early in the process can in fact render them more rather than less vulnerable to psychological distress" (Mutton, 2005).

As mentioned previously, clarity of communication channels and reporting arrangements within and between agencies is critical. There is a balance to be achieved between reporting to one's parent agency and meeting the lead agency's requirements, and with having the ability and necessity to make decisions on behalf of one's own agency staff offshore.

CONCLUSION

The role of Centrelink Social Workers in providing personal support to affected Australians offshore during disaster recovery is still developing. To the authors' knowledge, Australia is one of few countries to have activated such a response capability. This article has provided an overview of the service delivery framework and an exploration of the practice challenges, dilemmas and learnings. As mentioned previously, further elaboration of the Social Work knowledge and theoretical perspectives that informed practice was outside the scope of this paper. Given the ground-breaking nature of this practice and service delivery framework, the authors are committed to further exploration, documentation and continuous improvement.

REFERENCES

Austin, L. (Ed.) (1992). *Responding to disaster: A guide for mental health professionals.* Washington: American Psychiatric Press.

Canfield, J. (2005). Secondary traumatization, burnout, and vicarious traumatization: A review of the literature as it relates to therapists who treat trauma. *Smith College Studies in Social Work, 75*(2), 81-101.

Centrelink (1999). *Centrelink Annual Report 1998–1999*. Retrieved November 12, 2005, from http://www.centrelink.gov.au/internet/internet.nsf/filestores/ann_ rep9899/$file/05.htm

Centrelink (2000). *Centrelink Annual Report 1999–2000*. Retrieved November 12, 2005, from http://www.centrelink.gov.au/internet/internet.nsf/ar9900/chapter5.htm#ref4

Centrelink (2001). *Centrelink Annual Report 2000–2001*. Retrieved November 12, 2005, from http://www.centrelink.gov.au/internet/internet.nsf/filestores/ar0001pdf/$file/ar001chp5.pdf

Centrelink (2002). *Centrelink Annual Report 2001–2002*. Retrieved November 12, 2005, from http://www.centrelink.gov.au/internet/internet.nsf/ar0102/7_2.htm

Centrelink (2003). *Centrelink Annual Report 2002–2003*. Retrieved November 12, 2005, from http://www.centrelink.gov.au/internet/internet.nsf/ar0203/6_4.htm

Centrelink (2004). *Centrelink Annual Report 2003–2004*. Retrieved November 12, 2005, from http://www.centrelink.gov.au/internet/internet.nsf/ar0304/4_3_02.htm

Centrelink (2005a). *Centrelink Annual Report 2004–2005*. Retrieved November 19, 2005, from http://www.centrelink.gov.au/internet/internet.nsf/filestores/ar0405pdf/$file/chapter_5.pdf

Centrelink (2005b, January). On the ground services. *Centrelink People Special Edition, 33*. Canberra: Author.

Compton, B., & Galaway, B (1999). *Social work processes* (6th ed.). New York: Brooks/Cole.

Cowger, C.D. (1994). Assessing client strengths: Clinical assessment for client empowerment. *Social Work, 39*(3), 262- 268.

Dyregrov, A. (2001). Early intervention–A family perspective. *Advances in Mind–Body Medicine, 17*(3), 168-174.

Emergency Management Australia (2004). *Recovery*. Canberra: National Capital Printing.

Garvin, C.D., & Seabury, B.A. (1997). *Interpersonal practice in social work: Promoting competence and social justice.* (2nd ed.). Needham Heights, MA: Allyn & Bacon.

Gribbon, M., McLennan, L., Millar, S., & Newton, T. (2004). *The Bali support team: The development of a new service delivery model.* Unpublished paper to the IFSW World Congress Global Social Work, Adelaide.

Hodgkinson, P., & Stewart, M. (1991). *Coping with catastrophe*. London: Routledge.

Howe, David. (2003) *An introduction to social work theory*. Ashgate: Aldershot.

Management Advisory Committee (2004). Case studies: Response to Bali bombings. In *Connecting government: Whole of government responses to Australia's priority challenges* (appen. 2). Retrieved November 12, 2005, from http://www.apsc.gov.au/mac/connectinggovernmenta2bali.htm

Mutton, D. (2005, February). The post disaster industry in the wake of the tsunami. In *Perspective*. Retrieved November 12, 2005, from http://www.abc.net.au/rn/talks/perspective/stories/s1298578.htm

Payne, M. (2005). *Modern social work theory* (3rd ed.). Hampshire: Palgrave MacMillan.

doi:10.1300/J198v05n03_05

With the Best of Intentions: Singapore Volunteers and Donors and the Tsunami Reconstruction

Tay Keong Tan
Paige Johnson Tan

SUMMARY. Charities and humanitarian organisations in Singapore that were involved in post-Tsunami relief and reconstruction work are repeatedly queried by volunteers, donors, and journalists. *The world has pledged and donated billions to Tsunami relief and reconstruction, and I have personally given money to Singapore charities. Why do I still read about needy people in temporary shelters? Where has the money gone?* Another question reflects similar sentiments of puzzlement, even frustration: *I am moved by the devastation and would like to help by volunteering my time and expertise. I have signed up to volunteer with several humanitarian organisations. I am highly qualified and could help people. Why am I still not sent overseas?*

Dr. Tay Keong Tan, PhD, is Executive Director, Singapore International Foundation, Bishon Pl. Jct. 8 Level 9, Singapore 579837, an agency involved in Tsunami reconstruction in Sri Lanka and Indonesia. He serves as a member of the Tsunami Reconstruction Facilitation Committee responsible for coordinating the involvement of Singaporean non-governmental organizations in the reconstruction efforts.

Dr. Paige Johnson Tan, PhD, is Assistant Professor of Political Science specializing in Indonesian politics, University of North Carolina-Wilmington, 601 South College Road, Wilmington, NC 28403. She is currently on research leave in Southeast Asia.

[Haworth co-indexing entry note]: "With the Best of Intentions: Singapore Volunteers and Donors and the Tsunami Reconstruction." Tan, Tay Keong, and Paige Johnson Tan. Co-published simultaneously in *Journal of Social Work in Disability & Rehabilitation* (The Haworth Press, Inc.) Vol. 5, No. 3/4, 2006, pp. 97-108; and: *Asian Tsunami and Social Work Practice: Recovery and Rebuilding* (ed: Ngoh Tiong Tan, Allison Rowlands, and Francis K. O. Yuen) The Haworth Press, Inc., 2006, pp. 97-108. Single or multiple copies of this article are available for a fee from The Haworth Document Delivery Service [1-800-HAWORTH, 9:00 a.m. - 5:00 p.m. (EST). E-mail address: docdelivery@haworthpress.com].

Available online at http://jswdr.haworthpress.com

doi:10.1300/J198v05n03_06

These are good questions that deserve solid answers. This article discusses the latest thinking in relief circles as to which volunteers are best sent overseas and how donations can most responsibly be used for effective and sustainable reconstruction. The authors draw on Singapore's experience with providing relief in the aftermath of the Tsunami along with Dr. Tay Keong Tan's observations of relief and rehabilitation efforts in the Tsunami-affected areas of Sri Lanka and Indonesia. The article finds that volunteering requires much more than just a giving heart. Also, donations may be used most effectively by thinking long-term rather than just short-term. Further, donations are best administered in a way coordinated with local relief plans and in consultation with local people. The article closes by considering lessons for volunteers, donors, and humanitarian relief agencies from the Tsunami's aftermath. doi:10.1300/J198v05n03_06

KEYWORDS. Volunteer, disaster, relief, reconstruction, accountability, humanitarian, Tsunami

MANY HAVE CALLED, BUT FEW ARE CHOSEN

Why is it that giving, able, and highly competent individuals seeking to be volunteers have been overlooked in the Tsunami's wake? Are non-governmental organizations (NGOs) too busy, exclusive or incompetent to meet volunteer interests? The question, as it turns out, is a misguided one. It suggests that all volunteers are needed, beneficial and welcome in the aftermath of a major disaster. The right question to guide volunteer deployment should be: what resources and assistance are needed at what point and does an agency have the means to place them responsibly and safely in the field? The people in the humanitarian sector have seen this so often that we have a term for it: "supply-driven volunteer deployment," where volunteer agencies send volunteers overseas more as an outlet for the volunteers' aspirations "to do something" than to meet an assessed need in the field.

One of the most commonly cited blunders of well-meaning humanitarians is the supply-driven mismatch of volunteer skills with needs in the disaster sites. International volunteers rush in often without the requisite preparation and coordination. Some stories from the Tsunami zones show the mismatch: doctors queuing for patients at a hospital camp in Galle, Sri Lanka; unskilled volunteers doing non-essential (manual) work in Meulaboh, Aceh while unemployed locals looked on; and the motley international aid contingents clogging the airports and supply lines leading into disaster zones.

When humanitarian workers descended upon the Tsunami-affected eastern coastline of Sri Lanka, for instance, we saw in one of the internally displaced persons camps the meager belongings of a family of five. Their possessions would fit into a single large suitcase. As NGOs streamed there to "set up tents and fly the flag," we witnessed NGO leaders lining up at the District Administrators Office in Ampara to offer their aid to local officials. We could see on a flipchart the long list of international agencies offering aid in health care, water and sanitation, housing, and education–a public listing of the ironical reality of too many aid agencies chasing after too few projects. One international aid worker leaving the office said: "Look around you, there are so many untrained foreign reconstruction crews building temporary shelters when newly unemployed Sri Lankan men stand around watching; there must be a better way to rebuild this country!"

In Aceh, Director of the Rehabilitation and Reconstruction Agency (*Badan Rehabilitasi dan Reconstruksi,* BRR) Dr. Kuntoro Mangkusubroto lamented that during the immediate aftermath of the Tsunami and subsequent Nias earthquake, foreign donors sent noodles, skirts and blankets, even cement, when what people most needed were sarongs, tents, sanitary pads, and biscuits (crackers) (Mangkusubroto, 2005). Some *thirteen hundred* containers of donated materials just sat in Medan's nearby harbortown at one stage. Kuntoro further commented that people who rushed to the site of the disaster were not a help but became a burden as they needed to be housed, fed and given drinking water, items that were needed by Tsunami-affected persons as well.

The Lifecycle of a Disaster

Disaster management experts often refer to distinct and discernable stages in the life cycle of disasters, also broadly applicable to post-conflict situations–the emergency, relief, rehabilitation and reconstruction phases. The emergency phase is a time for scrambling–evacuations to save or preserve lives, fire- and flood-fighting, and search and rescue. Swift action is most critical to get help to the people to prevent or limit the eventual impact. Immediately following that is the relief phase, when urgent measures are taken to meet the most essential needs of the survivors by providing water, food, shelter and medical care. Relief work is soon transitioned into rehabilitation, as the actions of the affected communities and aid workers turn to restoring basic services and assisting survivors in building their lives and means of livelihood. The final phase following rehabilitation is the long protracted period for reconstruction, when damaged infrastructure is rebuilt and all essential services restored in the context of economic recovery and a long-term development plan.

In the Red Cross and Red Crescent's *World Disasters Report 2005,* an evaluation of post-Tsunami work in Aceh is offered which highlights a number of interesting findings (International Federation of Red Cross and Red Crescent Societies, 2005). "The highly "visible" health sector attracted the most agencies: 22 health NGOs were operating in one area on the west coast [of Aceh]. Ten international field hospitals were set up in Banda Aceh, none of which worked at full capacity. There were too many surgeons. One UN witness in Meulaboh saw "20 surgeons competing for a single patient. Yet midwives and nurses were in short supply" (International Federation of Red Cross and Red Crescent Societies, 2005, Chapter 4). In the early phases of the recovery effort, UN officials already began to warn of "Tsunami tourists"–non-essential humanitarian workers touring the disaster cites, surveying the devastation and taking photographs but contributing little else.

Placement of volunteers overseas in a developing country, especially in a disaster zone, presents many physical, psychological and social challenges. They range from adjusting to the local diet, potential housing and hygiene hardships, to coping with local stereotypes about and prejudices against foreigners. Especially in responding to complex emergencies where a natural disaster overlays a civil strife situation such as in both Sri Lanka and Aceh, the preparation of volunteers and coordination of their work must be carried out with the utmost care. Suzanne Charest of the Canadian Red Cross writes: "When you are talking about emergency disasters of this size and scope, it's something that you really need to think twice about before spontaneously going over there to offer to help. If you've no training and experience in dealing with traumatic situations, it probably isn't a very good idea. We never send spontaneous volunteers to do relief work overseas" (Izon, 2005).

Why this level of selectivity and caution? The service provided by international volunteers is typically more complicated than that performed by domestic Samaritans. The challenge of aiding recovery and development is a complex and continuous process. Thoughtful, sustainable projects empower people and communities to rise up on their feet and take control of their lives and livelihood. Their struggle with disaster and disadvantage will continue for years. In this context, reconstruction projects, to have a lasting impact on the living standards for many people, must have local ownership and participation, and they must contribute to a larger policy framework supported by the local governments and the wider development donor community. For instance, instead of supporting stand-alone projects, the development community is now encouraging aid agencies to work within nationally-driven reconstruction plans and sectoral frameworks that address poverty, conflicts, poor governance, scarce resources, and diseases. Before aid agencies build a school, we have to ensure that it will be used to deliver effective services to the local pupils and it will be well run and adequately staffed by local people long after the internationals leave.

The well-prepared and committed international volunteer can be a positive presence in a community, a source of inspiration and empowerment. But the outsider's role can also be disruptive and burdensome. There have been countless examples of outsiders who impose their cultural norms, force their own agenda, and initiate projects without the participation of the local community. The results can range from inappropriate to disruptive or destructive, and the local people are left to put the pieces back together long after the well-intentioned visitor has left. Overseas volunteers must bear in mind that much of the services that foreigners fly in to provide can also be more efficiently provided by locals at a fraction of the cost. The Chairman of the Singapore Red Cross commented that it was "a mistake" to send a first group of volunteers to Aceh during the immediate rescue phase, largely to meet the volunteers' strong pent-up interest to get to ground zero to offer help (Choo, 2005b). The volunteer teams were sent in the first quarter of 2005 to Meulaboh to perform physical work of clearing debris and erecting temporary structures (tasks that could be more efficiently carried out by Acehnese).

It was not for a lack of volunteers that the United Nations Development Programme (UNDP) and the US-based Mercy Corps have a special scheme to hire local people (especially former fishermen) to work on various projects in Banda Aceh. There is a concern that young Acehnese men will leave town and further weaken the affected communities' cultural fabric and the prospects for sustainable long-term recovery.

Whether to comfort the sick, help the needy, or raise money for charities in our local community, domestic volunteers typically perform a service in a familiar local context. Often, disaster-affected communities are recovering from shock, their support systems decimated, and the structures for their day-to-day survival threatened. Overseas aid workers, hence, must be prepared to take on challenges far beyond the simpler tasks of domestic volunteering. An extraordinary degree of tolerance and cultural sensitivity must be observed. Singapore military officers learned to take off their sunglasses when they spoke to locals in Aceh, and aid distributors found it helpful to have their rations clearly marked with *halal*-certified[1] signs. To make a positive impact on the lives and livelihoods of people, international volunteers must ensure that contributions are welcome and are effective in the cultural context. Beneficiary communities expect and deserve nothing less.

Hence, volunteers for overseas humanitarian and development missions can take heart in the fact that not all Samaritans should be deployed overseas. Local people can do much of the work, given time and the right resources and opportunities. And the humanitarian community has a responsibility to work with local peoples and support their initiatives to meet their needs and empower them to rise up to overcome disaster and disadvantage in a sustainable and culturally appropriate way. While volunteers may be fueled by passion and the strong urge to help, the very best of intentions, volunteer managers have to a take a dispassionate and calculated perspective of the reconstruction challenges ahead.

MUCH IS GIVEN, MUCH IS EXPECTED

Non-governmental organizations in Singapore are regularly posed two questions: one relating to volunteering overseas in disaster situations and the other concerned with how money donated to the Tsunami disaster has been used. Now, we turn to the issue of money for Tsunami relief and reconstruction. News reports continue to show displaced and needy people who are still suffering. How can that be, when so much money was donated? Are there funds "lost" to organisational overhead, inefficiencies, or worse, corruption?

The Tsunami was an unusual disaster in recent history because of the unprecedented international impact and global response. Twelve countries were affected, with about a quarter million killed within the span of just a few hours, offering very little lead time for evasive action and response. Some 8,000 kilometers of coastlines were devastated, and these included three conflict zones in Aceh, Sri Lanka and Somalia. This was truly an international disaster, with people from more than 35 nations affected in Phuket, Thailand alone (the Tsunami caused the greatest loss of life of any natural disaster in faraway Sweden's modern history). Flash appeals during the immediate aftermath saw billions of dollars pledged, with a sizable proportion from the private sector. There was extremely stirring media coverage, with dramatic images of the mayhem broadcasted straight into living rooms all around the world. By any yardstick, this was a disaster characterized by the unprecedented scale and speed of the international response. A UN official called it "the biggest relief operation launched by the Organisation" (OCHA Official, 2005).

Singaporeans poured out their hearts and dug into their pockets, and organisations of every stripe raised funds for disaster relief. In the ultimate count, some US$52.8 million was raised for the Asia Tidal Waves Fund in response to the Singapore Red Cross Society's public appeal.[2] The appeal specifically called for donations in cash as it is the most efficient means of providing relief assistance; the needed relief items could be purchased in bulk and then delivered and distributed quickly to the victims.

The international donor community, too, is concerned about the effective and accountable use of the money given for Tsunami reconstruction. In Indonesia, a Multi-Donor Trust Fund for Aceh and North Sumatra (MDTF) of US$500 million funds has been specially created to pool donor resources to finance projects and programs to assist in the reconstruction. The MDTF pools donor resources to support a portfolio of projects and programs within the Government's Master Plan for recovery in Aceh and Nias.[3] It also serves as a forum for donor coordination for the various projects to regenerate communities and community infrastructure, restore livelihoods and rebuild governance in Tsunami-affected areas. The MDTF helps to fund the work of the Indonesian government's *Badan Rehabilitasi dan Rekonstruksi* (BRR, 2005), the body established to oversee reconstruction and rehabilitation work in Aceh and Nias. The BRR holds ministerial level authority and has the capacity to eliminate bureaucratic or organisational laziness and undertake autonomous planning and delivery. For probity, BRR has established an Anti-graft Task Force with a budget of Rp 1 million (US$100,000) to monitor corruption.

Questions have been raised over international involvement in reconstruction efforts after previous natural disasters. For instance, US Agency for International Development assistance to El Salvador following an earthquake there in 2001 raised questions as to the safety of building materials and techniques used in the reconstruction of homes.

Risks exist for the post-Tsunami rebuilding as well–from the suitability of the design to the quality of materials used. Asbestos is a commonly used construction material in Sri Lanka, and building contractors resort to sub-quality materials in a practice called "reverse mark-ups" in Indonesia. Will projects be well carried out when the international spotlight has long since moved on to the next disaster or terrorist attack? *How do we guard against these reconstruction pitfalls?*

In Singapore, a new structure to oversee the use of funds raised from the public and to monitor the progress of various reconstruction projects was set up. The monies contributed by the public were consolidated in the Singapore Red Cross' Asia Tidal Waves Fund. To allocate and disburse the funds, a Tsunami Reconstruction Facilitation Committee (TRFC, a group of heads of civil society organisations) was created to receive, assess and award project funds to civil society groups. The Committee was set up in January 2005 to help coordinate the people sector's involvement in the reconstruction efforts in the Tsunami-affected countries by evaluating project proposals to tap on the money donated to the Singapore Red Cross. Apart from people sector projects, the Committee also works with other agencies in supporting key reconstruction projects, such as the building of a barge-pier, a hospital and schools in Aceh and the construction of several schools and a clinic in Sri Lanka.

The committee reports progress in the use of the public donations for the Tsunami reconstruction projects by providing information on the approved projects, by country, sector and amount approved for the project. To this writing, projects by the Singapore Red Cross, Mercy Relief, Habitat for Humanity, YMCA, Central and Southwest CDC, and the Singapore International Foundation, along with others, have been approved by the TRFC. The repertoire of projects reflects the broad representation of civic, corporate, religious and special interest organisations in Singapore, and their work focuses on diverse sectors, ranging from community development and housing to health, education and economic recovery. The Red Cross Society's Website[4] lists the funded projects and regularly reports progress in the projects.

To ensure that the funds are effectively and efficiently allocated, the projects have to undergo strict eligibility requirements for funding evaluated by the TRFC. The Committee engaged Integrated Project Systems (IPS), an international consulting firm, to develop a tracking, monitoring and reporting system for the disbursement of funds for projects.

As of September 2005, close to US$27.6 million, or over 50% of the donated money, had been allocated to reconstruction efforts in Tsunami-affected areas (Choo, 2005a). Most was approved for Aceh (65%), followed by Sri Lanka (30%), and the Maldives (5%). This leaves only about US$25.2 million for new reconstruction projects as we enter the height of the reconstruction phase in late 2005 and 2006. For those who ask: Why is there still so much money undisbursed, there is actually precious little left for the remaining reconstruction work just one year into the job. Reconstruction is a very complex and long-drawn process that takes years. The BRR (established only in April 2005) has a mandate to work in Aceh and Nias for *four* years based on the assessed scale of infrastructure reconstruction and livelihood rehabilitation.

Hence, NGOs in Singapore must educate Singaporeans as to how their donated funds are being used and how a measured approach to disbursing funds is beneficial to a coordinated, consultative response to the disaster. Gradual and deliberate use of funds is a sign, not of laggardly administration, but rather of effective and accountable stewardship of public donations. Still, this is not an excuse for allowing red tape and inertia to slow down the stream of aid. Agencies responsible for rebuilding in Aceh were criticized for the slow pace of their efforts, as 65,000 people were still living in tents nine months from the disaster (Wake-up call for Aceh donors, 2005). In today's world, there are not many disaster mitigation and developmental challenges that require unplanned international interventions. Good management of donations must balance the race against time to shelter the homeless and feed the hungry with prudent allocation of money and careful planning and implementation of projects.

LEARNING FOR THE FUTURE

So what lessons has the Asian Tsunami offered to individual volunteers and donors as well as humanitarian agencies?

The first is that there is a time for swift action and rapid response; this is best carried out by organisations with strong and proven emergency response capacities such as the United Nations and the Red Cross/Red Crescent. In addition, the Tsunami showed that military involvement can be of great use; this includes the militaries of the affected nations as well as those countries invited to assist them.

Other organisations and individuals trooping in immediately after a disaster merely stretch existing resources. Although well-intentioned, these volunteers might be more of a burden than a blessing in the early stages. It is important to note that "crowding in" during the early phase of the disaster was not just the preserve of "Tsunami tourists." Respected international NGOs were also guilty of a "compulsive desire to do good" in the Tsunami's wake.

While the impetus to get out in the field and help feels most urgent immediately following a disaster (technology enables images of suffering and destruction from around the world to be broadcast into our living rooms), aid agencies, governments, international organisations, and media can help the long-term prospects of those in disaster zones by engaging in public education of the world's citizens. Rather than making a one-off cathartic donation, citizens should be encouraged to see disaster recovery as a long-term process; likewise, while fatiguing, citizen interest and engagement with the tragedy should be long-term as well. Donors should be overseers of the funds they provide, remaining interested throughout the life cycle of the tragedy to ensure that monies pledged are well-used. If Acehnese are still living in temporary shelters almost one year after the disaster, citizens must be encouraged to remain aware of their conditions and to let relevant governments, international agencies, and non-governmental organisations know that they are aware. Hold relief providers' feet to the fire to ensure that those in need are not forgotten. Sustained media interest can be of invaluable assistance in keeping public attention on a particular crisis. Unfortunately, the media often quickly move on to the next big thing.

Detailed needs assessments, coordination with the affected governments, and consultations with citizens in disaster zones can be conducted as humanitarian agencies gear up to participate in the longer-term processes of relief, reconstruction, and recovery. Analysis and consultation, reflection and planning, when the natural impulse is to get into the trenches and help, require great discipline. However, assessments, coordination, and consultation ensure that aid is well-timed, appropriate in nature and kind, and in keeping with the desires, cultural norms, and absorptive capacities of local governments and disaster-affected persons.

Many in the aid community are used to working in an environment of resource shortages. One can, it seems, always use more money to offer better health care, more schoolhouses, and better irrigation systems. However, in the wake of the Tsunami (and perhaps as will be the case with other "telegenic disasters"), overall resources were not a problem. Using the bonanza of resources effectively and in a coordinated way was much more at issue. Stories of too many organisations on the ground, dividing up territory between them and competing to fly their own flags, dogged the Tsunami's aftermath. Coordination and consultation take time. Surely, lessons can be learned from delays in the Indonesian government's planning process to make for more expeditious results. But, slow planning is invariably better than no planning.

We must inculcate in the humanitarian community a more holistic sense of its accountability, helping Samaritans to depart from the "savior mentality." Typically, humanitarian agencies are responsive to the needs and interests of their funders, be they individuals, governments, or foundations. However, it is especially clear from the Tsunami that other stakeholders are owed accountability as well. Local governments and disaster-affected persons should not be passive recipients of (often white, Western) largesse. Those in the disaster zones, those most affected, are those with the greatest stake in how recovery takes place. They should be "subjects" of the reconstruction of their areas and not just "objects." They have resources, skills, and preferences to be mobilized in their own service. They should be consulted as to how funds should be spent. Similarly, governments, international organisations, and NGOs should also account, through public forums to affected persons, for how money and resources raised in their names were used.

Lastly, there is no one right way to provide volunteers or donations for disaster relief. The balance between fast response and thorough planning, the importance of consultation as opposed to quick results is still being worked out in the field, in government offices, and at the headquarters of humanitarian agencies. While the Sphere Project, grouping the International Red Cross with non-governmental organizations like Oxfam, attempted in 2004 to put together a Humanitarian Charter and Minimum Standards in Disaster Response, effectively a code of conduct for those operating in disaster zones, this standard is not well-known even among members of the humanitarian community (Sphere Project, 2004).[5] Nor is the standard binding in any way. At book length, the standards and accompanying indicators are also somewhat unwieldy, making their operationalization extremely involved. We are left to learn from experience and apply those best practices we observe.

A UN official from the Office for the Coordination of Humanitarian Affairs reflected: "Relief agencies are notorious for not learning from the experience of previous disasters. They tend to respond to each disaster as if it was the first time" (OCHA Official, 2005). Let us hope we do not lose the lessons from this last year. With all our advanced technology, humankind still cannot prevent some natural disasters, or even accurately predict them. But we can learn to be better prepared and to respond better when they hit. When we fail to learn from past disasters and take their lessons to heart, then lives may be lost and suffering prolonged needlessly. This would be an even greater catastrophe.

NOTES

1. *Halal* indicates food that is religiously permissible for Muslims to eat, along the lines of kosher in the Jewish faith. This marking is vital in Muslim areas, particularly among the pious Acehnese.

2. This total represents about US$12 for every member of Singapore's population of 4.2 million. For population data, see Statistics Singapore at http://www.singstat.gov.sg/keystats/people.html#demo

3. The Master Plan may be found online at the Indonesian government's Aceh reconstruction information clearinghouse website at http://e-aceh-nias.org/index.php?id=master_plan.html

4. See the Red Cross website at http://www.redcross.org.sg/

5. Information on the charter and standards can also be found online at http://www.sphereproject.org/

REFERENCES

BRR (2005). *Aceh masterplan.* Retrieved March 9, 2006, from http://e-aceh-nias.org/index.php?id=master_plan.html

Choo, W. (2005a). How has Singapore helped restore lives of the people in the tsunami affected countries, nine months after the disaster? (press release). September 23, 2005.

Choo, W. (2005b). Remarks at the Singapore Humanitarian Conference, Singapore, August 2005.

International Federation of Red Cross and Red Crescent Societies (2005). *World disasters report 2005.* Retrieved March 9, 2006, from http://www.ifrc.org/publicat/wdr2005/index.asp

Izon, L. (2005). Volunteer vacations target the tsunami zone. *Globe and Mail* (Canada), January 15, 2005.

Mangkusubroto, K. (2005). The route to a modern Indonesia: Rebuilding from zero (public lecture). Singapore, August 31, 2005.

OCHA Official (Official with the Office for the Coordination of Humanitarian Affairs, UN) (2005). Remarks at the Enhancing the capacity and role of NGO networks in disaster: A workshop by the Asian disaster reduction & response network (ADRRN), Chennai, India, August 2005.

Sphere Project (2004). *Humanitarian charter and minimum standards in disaster response.* Oxford: Oxfam Publishing.

Wake-up call for Aceh donors (2005). *Straits Times* (Singapore), October 5, 2005, 13.

doi:10.1300/J198v05n03_06

Training for Disaster Recovery: A Review of Training Programs for Social Workers After the Tsunami

Allison Rowlands

SUMMARY. This paper describes a range of training programs for social workers and other recovery workers following the Indian Ocean Tsunami of December 2004. These programs were developed and implemented by the author in Singapore, and with collaboration from Indonesian colleagues, in Indonesia. The content is outlined and the rationale behind the development of the programs is presented. The theoretical bases for the diversity of interventions are argued. A course module for both undergraduate and postgraduate social work education is also described, as inclusion of crisis and disaster recovery management in professional courses is necessary to prepare practitioners for their inevitable involvement in responding to emergencies. doi:10.1300/J198v05n03_07

KEYWORDS. Training, social work, disaster recovery, Tsunami, psychosocial

---

Allison Rowlands, PhD, is Business Manager Social Work, Centrelink Area Hunter, 12 Shannon Street, Marks Point, NSW 2280, Australia (E-mail: allison.rowlands@centrelink.gov.au).

[Haworth co-indexing entry note]: "Training for Disaster Recovery: A Review of Training Programs for Social Workers After the Tsunami." Rowlands, Allison. Co-published simultaneously in *Journal of Social Work in Disability & Rehabilitation* (The Haworth Press, Inc.) Vol. 5, No. 3/4, 2006, pp. 109-126; and: *Asian Tsunami and Social Work Practice: Recovery and Rebuilding* (ed: Ngoh Tiong Tan, Allison Rowlands, and Francis K. O. Yuen) The Haworth Press, Inc., 2006, pp. 109-126. Single or multiple copies of this article are available for a fee from The Haworth Document Delivery Service [1-800-HAWORTH, 9:00 a.m. - 5:00 p.m. (EST). E-mail address: docdelivery@haworthpress.com].

Available online at http://jswdr.haworthpress.com

doi:10.1300/J198v05n03_07

INTRODUCTION

Critical to effective response in emergency situations are the skills and knowledge of those workers who provide a range of services at all stages of the recovery. Social workers are equipped through their core professional training to assess, provide crisis intervention, and deliver a range of counselling and bereavement services to individuals, couples, families and small groups (Cowles, 2003; Rowlands, 2004a). They are also equipped, through training in community development and social policy development, to intervene at a broader community or societal level and thus possess the capacity to contribute systemically to an affected community's recovery (Emergency Management Australia, 2003; Ife, 2002; Rowlands, 1996). Additionally, research is a mode of social work intervention, whether it is action research or through the contribution to the knowledge base, and serves as an important function in disseminating practice experience in disaster recovery (Emergency Management Australia, 2002; Padgett, 2002). It is also acknowledged that trained mental health workers, including social workers, experienced in supervision and management of teams, are well-suited to provide these essential services in emergency situations, and to support the wide range of workers in the recovery operation (Emergency Management Australia, 2002; Rowlands, 2004a). So why is it then that many colleagues do not feel well-prepared for this site of social work practice?

This article describes a range of training interventions planned and delivered by the author after the Asian Tsunami of 26 December 2004. A range of training models and content is proposed, tailored to the specific roles trainees are expected to assume, and the stage of recovery. Delivery of training after impact is problematic, yet it is argued that to ensure quality service provision and confident, competent practitioners, some level of on-site training is usually warranted (Raphael, 1986). Additionally, some training input is valuable for social workers who may not be called upon to provide direct intervention to survivors but will inevitably intervene with people affected in some indirect ways by a major disaster. Training inputs of this nature are also described. At the next level of preparation, course content for both undergraduate and postgraduate social work education is presented. Such content has not commonly been considered core training in a social work degree program, yet given the frequency of both natural and human-made disasters in recent years, is becoming imperative. Examples of these courses developed by the author are presented. Finally, this paper also articulates a theoretical perspective that is seen as foundational in the preparation of social workers who will work in post-disaster or post-conflict situations. It argues that explicit adoption of a strengths-based orientation and a human rights framework, provides the firm foundation for effective intervention.

The purpose of this article is to describe some training programs that have been delivered and share some of the thinking behind these programs, so that future practitioners might be able to extend and develop some of the key ideas. It is acknowledged that a multitude of training packages exists and an exhaustive search for these has not been undertaken. However, by disseminating these ideas it is hoped that in future disasters some time may be saved in applying some of the ideas. It is further acknowledged that there is a wealth of material already available in related context areas such as trauma intervention–working with survivors of sexual assault and interpersonal violence being some obvious examples. Again, a rigorous search of such training materials was not undertaken in the time available. The final limitation of this article is that of "site"–the author was based in Singapore at the time of the Tsunami so the training material and delivery was limited to Singapore and Indonesian sites and does not attempt to address training delivered beyond the author's scope.

The author's previous experience in disaster recovery management in Australia, and in consulting to a psychosocial rehabilitation program in the post-conflict situation of East Timor in 2000, provided relevant experiences on which to build these training interventions. For example, in East Timor, a manual for Timorese staff in providing mental health interventions during the period of the United Nations Transitional Administration in East Timor (UNTAET) in 2000, was heavily edited by the author. It included contributions from several Australian social workers, including the author, over a six-month period in 2000 (PRADET, 2001; Rowlands, 2003).

THE POST-IMPACT SITUATION

Within the first fortnight after the Tsunami, the Social Work Department at the National University of Singapore (NUS) had reached out to colleagues in Indonesia, and the genesis of the FAST Project (2006) (Families and Survivors of Tsunami Project; for detailed information see the project Website at www.fastproject.net/) had been formed by Dr. Tan, Singaporean social workers and the author. As information about the devastation in Aceh became known, a fact-finding and consultation visit to Jakarta and Aceh was arranged by Dr. Tan and the author. This visit provided a springboard to plan future training of Indonesian social workers and staff of the Ministry of Social Affairs. It also enabled consultation regarding projects that could receive some funding from FAST in the future, and discussions with Ministry, academic and non-government organizations (NGOs), primarily about the needs of children and families who had survived and were now trying to rebuild their lives.

Colleagues within the Indonesian Association of Professional Social Workers (IPSPI), and the social work academic fraternity in Jakarta, welcomed input into a range of training initiatives for social workers in Jakarta itself, and those identified within the profession and the Ministry who would be delivering services in Aceh and northern Sumatra. Train-the-Trainer models were favoured for delivering training and collaborations with the Ministry, universities, NGOs and FAST developed. The focus of the training was community-based psychosocial management, recovery and rehabilitation services (Pujiono, 2005).

Content

The author prepared an outline for training content considered to be foundational to prepare workers anticipating a direct service role. A fundamental consideration in such program development is cultural sensitivity. Therefore, while a Caucasian social work academic may have the knowledge and training experience to develop material, it is crucial that colleagues from the culture and community where the training and the intervention will be delivered are involved from the outset in planning content, process and the consideration of values. The program as developed was delivered by Indonesian social workers and social work educators, for cultural, linguistic, ethical and values reasons. Training was delivered to varied groups of trainees at different times over the ensuing months, drawing from the menu of topics in the broad curriculum, and tailoring the content to the specific audience. As a consequence of this embryonic training intervention, funding obtained through the NGO Building Bridges to the Future has enabled ongoing training and support for the recovery process to rebuild the village of Rumpet. Rumpet is located in the Aceh province of Indonesia, which the World Heath Organization has deemed the worst hit by the Tsunami of 2004 (Building Bridges to the Future, 2006).

The program outline included crisis and trauma interventions, principles of disaster recovery management, loss and grief work, community-based methodology and community recovery intervention, cultural issues, values and ethics, debriefing, supervision and self-care (see Appendix 1). The proposal stated that the ultimate beneficiaries of the training project would be the "thousands of survivors and communities affected by the tsunami disaster who will receive the added-value in the goods and services that are delivered by direct service providers from government, international community and NGO sectors. The other possible direct beneficiaries are social work profession practitioners and scholars both in Indonesia and the ASEAN region" (Pujiono, 2005).

The importance of delivering this Train-the-Trainer model was underscored by the projections, in early 2005, of 160 social work and psychology trained staff being located in Child Centres in Aceh, each centre covering 20 camps. It was anticipated that these staff would be a combination of Ministry staff, newer appointments with little practice experience, and other social workers and psychologists including lecturers from the various institutions (Tan & Rowlands, 2005).

TRAINING IN SINGAPORE

Singapore offered considerable resources to devastated regions of several countries immediately. Notably, the Singapore Defence Services swung into action in Aceh and a major non-government collaboration of agencies coordinated volunteer-driven responses in Aceh and Sri Lanka. The Ministry of Health support included provision of hospital and medical teams to Aceh and Sri Lanka.

Volunteer Preparation in the NGO Context

A key concern of the NGO response was the preparation, training and debriefing of volunteers who would be providing a wide range of emergency and longer-term rebuilding programs as well as health, psychosocial and community redevelopment activities. As a result, the Young Men's Christian Association of Singapore (YMCA), in concert with other NGOs coordinating the volunteer recruitment activity, arranged short courses for prospective volunteers. Several of these brief workshops were delivered in January and February 2005 under the auspices of the YMCA.

Content

The author's content for one such session is in Appendix 2. One of the issues that needed addressing was the target audience itself. This was an unknown factor and potentially included a diverse group of people. Material needed to be presented in an accessible fashion, yet be of sufficient rigour for the participants to gain some knowledge and, more importantly, to facilitate self-reflection as to whether this was an appropriate undertaking for them. Thus, an objective was to inform participants of the current state of knowledge of human behaviour in disasters, consequent needs of communities, brief information about the kinds of interventions considered appropriate, and the impacts for workers/volunteers in such environments. Consequently, a focus on values was integral as well as discussion on cross-cultural work. Even though the majority of the audience were Singaporean Chinese, there is a cultural gap between their experiences and the lives of the Tsunami survivors in Aceh and Sri Lanka. Additionally, there were expatriate workers in the audience from America and Europe. Facts about the recovery environment and likely tasks of volunteers were addressed by the YMCA representatives. An overview of the situation in Sri Lanka, as currently known, was provided by a representative from the Embassy in Singapore.

Training for Professional Workers

A separate full-day training workshop was delivered by the author for social workers and allied mental health workers such as counsellors, mental health nurses or youth workers, under the auspices of the Singapore Association of Social Workers, in February 2005. Interest and anticipation had been building within the social work community in Singapore since the Tsunami occurred. Some were preparing to travel to the affected locations, mainly Aceh and Sri Lanka, to provide professional services under a diverse range of auspices. Some were keen to offer their services but did not have any structure and their enthusiasm was rather unfocused. Others were anticipating activation through governmental channels such as a hospital team. Still others were anticipating that they would be working within Singapore with citizens impacted by the Tsunami as survivors or because they were linked in some way with survivors or returning relief workers.

Content

The content of the third training program (see Appendix 3) is clearly linked to the previously described ones, with emphases varying according to the audience level of prior training, workshop objective, and format and time available. Thirty-one participants attended this workshop. The workshop used video trigger material and case studies to apply the theoretical material and stimulate discussion. Small groups worked on exercises in intervention planning and the workshop was experiential to a certain extent.

Evaluation of the workshop yielded the following data:

Content of training	good to excellent 68%	average 32%
Handouts and/or reading materials	good to excellent 86%	average 12%
Exercises, cases and/or role plays	good to excellent 47%	average 46%
Recommend this course to others	yes 92%	no 8%

Participants found the theoretical material on models of intervention helpful as well as case examples and discussions in small groups developing interventions. Particularly valuable were the sections on values and ethics, issues for workers, and interventions other than direct casework/counselling approaches.

Information Dissemination Through the FAST Website

A further mode of resourcing social workers and others delivering services to the affected region was through the FAST website itself. While the scope and nature of its audience is not known, it was clear through the offers to assist in the recovery process that this was an international audience and harnessed the imagination and commitment of many social workers. In the first three months post-Tsunami there were 25 offers of assistance from around the world registered through the website. There were also donations from individual social workers to the FAST Trust Account to provide seed funding for recovery projects (FAST Project, 2006).

In order to support the virtual or actual social work volunteers planning to do Tsunami relief work, or other related emergency recovery work, the author provided on-the-site reference material on many of the content areas from the training programs. It is reiterated that an overwhelming amount of valuable material is already available on the internet, yet for many "accidental" disaster recovery workers, being equipped with accessible, current, evidence-based and succinct reading material is of great value. To save people scrambling through search engines, a set of reading material was provided directly or linked to the

FAST site, as well as annotated bibliographies, to facilitate the acquisition of information.

Social Work Education in the Tertiary Sector: Undergraduate and Postgraduate

While the author had previously delivered a subject for masters level students in Australia on trauma and disaster recovery practice, such a subject had not previously been offered at NUS. As a result of the Tsunami, and in recognition of Singapore social work's prior experiences in emergency recovery situations (the SARS epidemic, Bali Bombings, Singapore Airlines and Silk Air plane crashes), the Department wished to mount a course module in crisis and disaster recovery management. The author developed this module and it was delivered for the first time for Honours undergraduate students in 2006. A modified course was also delivered by the author to masters level students in 2006. At the time of writing, the semester is still running, therefore no evaluation data is yet available on these two modules.

Content

For an outline of the content see Appendix 4. The rationale for the content as developed, was informed by a range of factors. Obviously the prior training programs for social workers and volunteers described above were a contributing factor. However, developing curriculum for academic study clearly demands a theoretical foundation and rigour in the content and pacing or structure of the course, as well as clearly articulated learning goals and assessment tasks. It was decided to offer the undergraduate module at Honours level students–those in their fourth year of social work education in the Singapore context. This guaranteed a prerequisite level of training and theoretical understanding, leaving the course the freedom to build on existing knowledge and focus on more specialized content. An experience-based, self-directed approach was favoured for pedagogical reasons (Gray & Gibbons, 2002) as well as the immediacy of the Tsunami event and subsequent emergencies internationally throughout 2005. These included Hurricane Katrina, the north Indian-Pakistan Earthquake, the first anniversary of the Tsunami, and the spread of avian flu, which resonated in a meaningful way for Singaporeans after their SARS experience in 2003.

The class size was 25 students which facilitated the use of groupwork techniques in skills and problem-solving activities, discussion of the material, and student group presentations. The class met for a three-hour session once weekly. The first weeks were resource sessions where the author presented theoretical material for discussion and critique after a trigger was presented. Heavy use was made of both video footage and case studies of the Tsunami as well as other recent disasters such as Hurricane Katrina, the Pakistan Earthquake and the Bali Bombings. One session was dominated by the viewing and subsequent discussion of the film "Into the Arms of Strangers: Stories of the Kindertransport" which reviewed the Kindertransport Program of early 1939 when 10,000 children, almost exclusively Jewish, were placed with foster parents and hostels in Great Britain and thus survived the Holocaust (Oppenheimer, 2000). The film incorporates interviews of several of these Jewish children, now adults. It thus formed a powerful vehicle for considering the theoretical material relating to loss, trauma, violence, war, genocide and family relationships. It also highlighted resilience, affording the reflection on strengths perspectives and growth through tragedy. The case vignettes that these films presented could then later be drawn upon to further elucidate theoretical points and for role play skills practice (from the disaster video material). Other video footage used included the Free Enterprise ferry disaster in the English Channel, the recovery operation after the Lockerbie plane crash, and a documentary about the Indonesian invasion and oppressive occupation and violence in East Timor, 1975-1999. "Bitter Paradise" (Briere, 1996) served as a trigger to consider the literature on state violence and interventions in post-conflict situations. The author's personal experience in training Timorese health workers and consulting to a fledgling community mental health service in East Timor in 2000/2001 was drawn upon to illustrate interventions and integrate theory with practice (Rowlands, 2003). A further resource session was a presentation of the author's recent research on the impact of SARS on Medical Social Work practice in Singapore (Rowlands, 2004b), used to illustrate the application of trauma and disaster theory to a real, local example with which the students could easily identify. Issues pertaining to training, intervention models used, the impact of the epidemic on professional and personal life, and workers' coping strategies, were highlighted.

The students' presentations in small groups had specific learning objectives and assessment expectations (see Appendix 4). They presented as a case example a particular disaster, developed intervention plans addressing the needs of at least two different groups affected by the disaster, and were to critique the theoretical material. They were to make links between the case example, recovery interventions and the local context, providing recommendations for how such an emergency should be handled locally. Remaining assessment comprised an individual written paper and an examination. Students chose a diverse range of disasters including the Singapore Airlines plane crash in Taipei in 2000, the collapse of the New World Hotel in Singapore in 1986, killing 33 people, the Columbine High School shooting in 1999, and the London bus/train bombings in 2005.

Using an experience-based framework for the learning process was both apposite and risky, from the author's point of view. Students acknowledging their emotional responses to the learning process will enhance self-awareness and provide a vital component in developing understanding of their future clients' reactions and experiences. It also provides opportunities for reflection and greater consolidation and integration of the learning being undertaken. Risks included students' own experiences of losses being evoked, especially when local incidents were discussed. However, one student group chose to develop their presentation around the Silk Air MI 185 plane crash in Indonesia in 1997 in which three secondary school teachers from their school (which they were currently attending) were killed. Students were able to share their feelings about the issues being discussed in a group climate of trust. It was important to contract at the beginning of the module regarding what was appropriate sharing within the class and confidentiality expectations. Nonetheless, the risk always exists of inappropriate disclosure of personal concerns. Current tertiary students in the Singapore context have experienced rather traditional learning approaches in their secondary education, hence experiential models are less familiar to them, while also having the advantage of being novel.

THEORETICAL FOUNDATIONS

The modules at both undergraduate and graduate levels drew explicitly on theories of crisis, loss and trauma. Intervention models were applied and critiqued for working with individuals, families and small groups, and communities. Principles of disaster recovery management, such as those espoused in manuals by Emergency Management Australia (1996), the International Federation of Red Cross and Red Crescent Societies (2006) and the Sphere Project (2004), provided current examples of the operationalisation of these theories in recovery planning and intervention.

Yet critical to effective recovery, and consistent with the above principles, is the explicit application of a strengths perspective (Saleebey, 2006) in responding to affected communities and facilitating rebuilding. As has been argued elsewhere, the capacity of individuals to recover and take charge of their personal and community rebuilding must be recognised by all authorities involved in the response (Rowlands, 2004a). It was therefore overt in the content of these modules that a considered appreciation of the literature from psychology and psychiatry, along with social work contributions in the areas of strengths/resilience and community development, is required. A review of the recent contributions in trauma literature already highlights the critical role of social support and other psychosocial care services (van Ommeren, Saxena, & Saraceno, 2005). Silove (2005) goes further in arguing that the best immediate therapy for acute stress is social. These authors are some of the many who now are attempting to redress an overly clinical, medical view in responding to communities impacted by disaster, and the privileging of psychopathology over normal coping responses to an abnormal situation.

The final tier to the theoretical foundation for the modules is that of a human rights perspective. Ife (2001a) argues that human rights form the basis of global citizenship, and are claimed by all people, regardless of race, culture, gender, age, or national origin. Social work as a profession is based on principles of human rights. Therefore, interventions in post-disaster or post-conflict situations must also be undertaken from within a human rights framework, as without that they will fail the individuals, families, communities and nations that they seek to serve. Because of their particular expertise, social workers, as human rights professionals, stand ready to work in partnership with governments and with NGOs in the pursuit of human rights (Ife, 2001b). As a case in point, Ife (1999) applied these principles to the international rebuilding efforts in East Timor during the UNTAET period. He cautioned against the re-emergence of colonialist statements and attitudes: "The good intentions, and the many achievements, of those working in NGOs and in the UN cannot be questioned, but the temptations of colonialism are subtle and insidious, and require a high level of personal awareness and political consciousness if they are to be resisted" (Ife, 1999). His assessment (from his perspective as Professor of Social Work at the time) was that "it would, however, be wrong to jump to the conclusion that therefore there is a need for trauma counsellors and extensive post-trauma programs to be designed and delivered from outside . . . we must also not underestimate the capacity of the East Timorese to deal with this themselves. Several East Timorese said to me that there are traditional methods of healing, which can be used at community level, and the very strong community structure of East Timor society means that there are very significant supports and capacities for this to be achieved through local community activity" (Ife, 1999).

CONCLUSION

In presenting an overview of some of the training programs initiated in Singapore and Indonesia after the Tsunami, the author has argued that training is a vital component of disaster recovery interventions. Training programs can be designed for very varied target groups who will be providing personal, psychosocial or community-based services in affected communities. While qualified professional mental health workers have a matured knowledge, values and skills base from which to draw for their practice, other recovery workers and volunteers may require foundational input on human responses to disaster and basic intervention techniques. Tertiary institution social work courses are also appropriate places to mount training in crisis and disaster recovery work and such a course is described.

Evaluation of training programs delivered is desirable to continue to improve the quality of training offered and to ensure that content is current and evidence-based. The sharing of course content and approaches is recommended to facilitate swift resourcing of our colleagues. Social workers will always be called upon to respond in disaster situations and it is hoped will become more integral to national recovery plans everywhere. Therefore, the preparation and training of social workers will indirectly facilitate the recovery of affected communities.

REFERENCES

Briere, E. (1996). *Bitter paradise: The sell-out of East Timor*. NY: Cinema Guild.

Building bridges to the future (2006). http://www.buildingbridgestothefuture.org/what.html. Accessed 2 March 2006.

Cowles, L.A. (2003). *Social work in the health field: A care perspective*. Binghamton, NY: The Haworth Press, Inc.

Emergency Management Australia (1996). *Australian emergency manual disaster recovery*. Canberra: Emergency Management Australia.

Emergency Management Australia (2002). *Mental health practitioners guide: Guidelines for psychological service practice.* Canberra: Emergency Management Australia.

Emergency Management Australia (2003). *Community development in recovery from disaster.* Canberra: Emergency Management Australia.

FAST Project (2006). http://www.fastproject.net/. Accessed 7 March 2006.

Gray, M., & Gibbons, J. (2002). Experience-based learning and its relevance to social work practice. *Australian Social Work*, 55(4), 279-291.

Ife, J. (1999). Report of assessment mission to East Timor. Report by Secretary, Human Rights Commission, presented to International Federation of Social Workers. http://www.ifsw.org/en/p38000240.html. Accessed 8 March 2006.

Ife, J. (2001a). *Human rights and social work: Towards rights-based practice.* Cambridge: Cambridge University Press.

Ife, J. (2001b). Commonwealth organisation for social work: Statement on human rights. Paper presented by Commonwealth Organisation for Social Work to Commonwealth Heads of Government Meeting, Brisbane, 2001.

Ife, J. (2002). *Community development: Community-based alternatives in an age of globalisatio*n (2nd ed.). Frenchs Forest, Australia: Pearson.

International Federation of Red Cross and Red Crescent Societies (2006). http://www.ifrc.org/what/disasters/IDRL/index.asp. Accessed 8 March 2006.

Oppenheimer, D. (2000). *Into the arms of strangers: Stories of the Kindertransport.* Burbank, CA: Warner Home Video.

Padgett, D.K. (2002). Social work research on disasters in the aftermath of the September 11 tragedy: Reflections from New York City. *Social Work Research*, 256(3), 185-191.

PRADET (2001). *PRADET staff training manual.* Dili: Psychosocial Rehabilitation and Development East Timor, unpublished document.

Pujiono, P. (2005). Proposal: Social work psychosocial TOT in Tsunami recovery and rehabilitation. http://www.fastproject.net/. Accessed 7 March 2006.

Raphael, B. (1986). *When disaster strikes.* New York: Basic Books.

Rowlands, A. (1996). Empowering communities to recover from disasters–A perspective on social support. Australian Red Cross NSW Disaster Services Seminar, Newcastle; published in proceedings.

Rowlands, A. (2003). Collaborative practice across borders: Experiences of Australian social work practice in East Timor, 2000-2001. *Asia Pacific Journal of Social Work* 13(2), 6-25.

Rowlands, A. (2004a). Reappraising social work's contribution to recovery from disaster and trauma: Applying a strengths perspective. *Asia Pacific Journal of Social Work and Social Development*, 14(2), 67-86.

Rowlands, A. (2004b). Hospital social workers' response to SARS in Singapore, 2003–Lessons learned from practice. Paper presented at the International Federation of Social Workers Congress, October 2004, Adelaide.

Saleebey, D. (Ed.) (2006). *The strengths perspective in social work practice* (4th ed.). Boston: Pearson Education.

Silove, D. (2005). The best immediate therapy for acute stress is social. *Bulletin of World Health Organization*, 83, 75-76.

Sphere Project (2004). *Humanitarian charter and minimum standards in disaster response*. Geneva: Sphere Project. http://www.sphereproject.org/handbook/index.htmj. Accessed 3 March 2006.

Tan, N.T., & Rowlands, A. (2005). FAST project, Report of visit to Indonesia 27 Jan - 1 Feb 2005. http://www.fastproject.net/. Accessed 7 March 2006.

van Ommeren, M., Saxena, S., & Saraceno, B. (2005). Mental and social health during and after acute emergencies: Emerging consensus? *Bulletin of World Health Organization*, 83, 71-75.

doi:10.1300/J198v05n03_07

APPENDIX 1. TRAINING OF TRAINERS FOR DISASTER MANAGEMENT AND TRAUMA COUNSELLING, IPSPI, INDONESIA

3 days from July 27-29 2005 at Medan Ministry of Social Affairs Training Centre.

COURSE CURRICULUM for "Overview of Disaster Management"

1. The 3-day training program is part of 12-day intensive Disaster Management and Trauma Counselling Workshop
2. The topics include:

(a) Overview/Principles of disaster management

(b) Crisis and trauma interventions–definition, theories, interventions, loss and grief, PTSD

(c) Trauma counselling

(d) Case management of stress, crisis and trauma, restoration and rehabilitation

(e) Social work in the context of disaster and displaced people–ground procedures and process

(f) Community-based methodology, community recovery intervention, and role of information–theories and strategies

(g) Tasks, roles and skills such as facilitation skills of recovery workers

(h) Debriefing and self-care as well as worker needs and supervision

(i) Cultural issues, values and ethics in disaster management and trauma counseling

APPENDIX 2. TRAINING FOR VOLUNTEERS, YMCA WORKSHOP, SINGAPORE, JANUARY 2005

Understanding sociocultural context and developing sensitivity
Worldview of volunteer–personal motivation, background, training
Identifying trauma, grief and loss–nonverbal and verbal listening, responding skills
Possible scenarios and tentative responses
Information and referral resources–in country–Sri Lanka update
Support for volunteers

- Understanding disaster
- Principles of recovery
- Phases of disaster
- Impacts of disasters–feelings; impacts on different groups
- Impacts on workers and self-care; debriefing
- Case example

APPENDIX 3. SPECIAL WORKSHOP ON DISASTER AND TRAUMA RECOVERY– SINGAPORE ASSOCIATION OF SOCIAL WORKERS/ FAMILY RESOURCE AND TRAINING CENTRE

Learning Objectives:

- To understand the psychosocial issues and needs for children and families after disaster
- To revise/develop core skills in crisis intervention and responding to loss
- To develop capacity to undertake a range of assessments after disaster
- To learn skills in responding to trauma reactions
- To understand the needs of staff and be able to provide a basic level of supervision and support for them

Content:

- Principles of disaster recovery intervention
- Effects of disaster on individuals, families, communities

- Assessment issues
- Crisis and trauma interventions
- Children
- Community recovery intervention and role of information
- Values and ethics of intervention
- Worker needs, supervision and debriefing

APPENDIX 4. SW4219 CRISIS AND DISASTER RECOVERY MANAGEMENT, DEPARTMENT OF SOCIAL WORK, NATIONAL UNIVERSITY OF SINGAPORE

Rationale:

Social workers have a role in responding to crisis situations, emergencies and natural disasters, both nationally and within the region. This module covers the theoretical and skills base to intervene effectively at the individual, family, group and community level, as well as the principles of international recovery management. The values and principles of recovery management are effectively those of community development, and examples of specific projects will also be explored. The module also addresses issues of project management, evaluation, staff supervision and debriefing, and emergency funds management.

Aims and Objectives:

- Understanding of reactions to disaster, trauma responses, individual and community needs
- Skills in crisis intervention and grief counselling
- Understanding of the context of disaster recovery management, recovery principles and politics, and interagency coordination
- Knowledge of project management, staff needs and supervision

Teaching Mode:

This module will be seminar-style and adopts an experienced-based approach. The class size is kept small to facilitate active participation. Group and individual exercises, role play and group discussion will be used rather than lecture format. Students will be expected to read for each class and be prepared to discuss their reading.

Syllabus:

- Crisis intervention and grief work
- Responding to trauma and disaster
- Case Example: The Role of Medical Social Workers in Singapore during the SARS Epidemic
- Case Example and Critique: Indian Ocean Tsunami
- Anatomy of disaster; planning for recovery
- Disaster Recovery Planning and Intervention at the individual and family level
- Disaster Recovery Planning and Intervention at the community level
- Caring for workers; evaluation and research
- Post-conflict situations, refugees and survivors of torture

Assessment
Group assessment task:

In small groups of 4-5 students, select a major incident (traumatic event, natural or man-made disaster, terrorist act, etc.). Develop an intervention plan that addresses both direct and indirect services to some of the affected population. Provide a critique of the theories or models upon which your intervention plan is based.

Facilitate a seminar presentation of about 1.5 hour's length for the whole class on your agreed topic. You may use any format and process for the seminar that you feel best communicates your learning to your colleagues and engages their minds in the learning process.

Individual paper:

Choose a specific focus from the range of social work interventions and tasks in facilitating individual and community recovery after a disaster or traumatic event. You may choose to link this to some part of the intervention plan you proposed in your group seminar, or choose a different topic altogether. Explore this topic in depth from the literature and discuss how you would adopt/adapt such an intervention or activity for the local context. Your paper should be theoretically informed and critical.

Length: 3000 words or 12 pages double spaced, excluding references

Both assessment tasks will be marked on the following criteria:

- content, use of literature and theory
- organisation and substantiation of ideas and arguments
- structure, expression, engagement of class (presentation); accuracy (paper)
- adherence to time/length guidelines and referencing

Seminar participation and final exam comprise the rest of the assessment marks.

The Role of Social Workers and Social Service Delivery During Crisis Intervention for Tsunami Survivors: A Case Study of Thailand

Pattamaporn Busaspathumrong

SUMMARY. This paper seeks to understand the economic, social and psychological impacts on survivors in Thailand of the Asian Tsunami on 26 December 2004. The Tsunami disaster brought about great changes in the lives of survivors, the role of social workers and social service delivery. Problems were actively worked out between many parties resulting in greater collaborations between local, national and international organizations. Social workers worked in a collaborative manner with various professions in delivering crisis intervention. In the case of the Tsunami disaster, there is an emergence of the development of the professional role of social workers and other social service workers to respond to the urgent needs of the family and children survivors. doi:10.1300/J198v05n03_08

Pattamaporn Busaspathumrong, PhD, is on the Faculty of Liberal Arts, Asian University, 89 Moo 12, Highway 331, Huaiyai, Banglamung, Chon Buri 20260, Thailand.

[Haworth co-indexing entry note]: "The Role of Social Workers and Social Service Delivery During Crisis Intervention for Tsunami Survivors: A Case Study of Thailand." Busaspathumrong, Pattamaporn. Co-published simultaneously in *Journal of Social Work in Disability & Rehabilitation* (The Haworth Press, Inc.) Vol. 5, No. 3/4, 2006, pp. 127-137; and: *Asian Tsunami and Social Work Practice: Recovery and Rebuilding* (ed: Ngoh Tiong Tan, Allison Rowlands, and Francis K. O. Yuen) The Haworth Press, Inc., 2006, pp. 127-137. Single or multiple copies of this article are available for a fee from The Haworth Document Delivery Service [1-800-HAWORTH, 9:00 a.m. - 5:00 p.m. (EST). E-mail address: docdelivery@haworthpress.com].

Available online at http://jswdr.haworthpress.com

doi:10.1300/J198v05n03_08

KEYWORDS. Tsunami, social work, social policy, Thailand, social services

Thailand experienced a great loss when the killer tsunami's waves crashed onto its shores on the 26th December 2004. This tragic event has impacted the lives of survivors, both Thai and foreign, and fundamentally changed the role of social workers and the social service delivery system. A National Economic and Social Delivery Plan has brought changes in both the rural and urban regions in Thailand minimizing the negative psychological, sociocultural and economic impact. This plan included the mobilization of local and international levels of social work practice and social work delivery to provide assistance to the victims.

The purpose of this paper is to describe the different effects of relief work during the crisis. The combination of local social work organizations and international organizations during the urgent relief period from December 2004 to January 2005 can be considered as a social work process in a voluntary context. This period exemplifies social service delivery in the era of globalization in terms of cooperation among local, national and international organizations. The boundaries of the social service system expanded as a result. It is recommended that long-term planning and future prevention can focus on the development of educational technology for disaster prevention and rescue work.

ECONOMIC AND SOCIAL IMPACT OF THE TSUNAMI

Economic Impact of the Tsunami

The greatest problem during the crisis was job loss and unemployment. In the early stage, one out of 10 people in Phuket province lost their jobs as tourism-related businesses were greatly affected. The Social Security Office (SSO) reported that almost 100 businesses suspended operations and more were expected to do so over the coming months. A huge percentage of closures include hotels, tour companies and restaurants. The unemployment rate is expected to rise further as more businesses are expected to close down. Phuket province faced the worst consequences among those areas affected (MCOT News, 2005).

The southern provinces of Thailand along the Andaman Sea which were affected by the Tsunami included Ranong, Phang Nga, Phuket, Krabi, Trang and Satun. (See Table 1.) The United Nations Environmental Programme (UNEP) reported 5,393 dead, 3,062 missing and 8,457 injured (Manuta, Khrutmuang, & Lebel, 2005).

Access to assistance was a significant problem during the crisis. Some of the victims in these six provinces had no access to the help which poured into Thailand. These people included fishermen who lost their tools and boats, villagers who could not return to their land, migrant workers who lost their legal documents, and some transient entrepreneurs.

Social and Psychological Impacts on Tsunami Survivors

The sudden shock of having homes destroyed and witnessing deaths of family and friends was an unforgettable traumatic event for the survivors. It is reported by the Public Health Ministry of Thailand that 5,000 Tsunami survivors consulted psychiatrists to deal with the psychological impact. Although a small number of patients were hospitalized, some refused to travel close to the sea, and many experienced great stress, insomnia, fear, guilt and confusion.

The impact of the killer waves has remained until now, over one year on. Essentially, the needs of women and men, boys and girls were different. Such gender perspectives were considered during policy formulation and long-term service delivery. This included services for

TABLE 1. Tsunami-Affected Villages in Six Southern Provinces of Thailand

Province	Total Villages by the Sea	Affected Villages	Deaths and Missing Persons	Damage houses (Totally destroyed)	Damage boats	Damaged fishing tools	Damaging fishing farms
Krabi	83	18	35	414	793	8,560	231
Phang Nga	123	46	1,616	2,267	581	N/A	1,792
Phuket	38	18	7	42	132	392	925
Ranong	43	14	146	221	1,053	199	300
Trang	57	36	34	100	285	26,118	666
Satun	174	29	8	33	463	458	11,620
Total	418	161	1,846	3,077	3,307	35,727	15,534

Source: CODI, 2005 (p. 32).

economic recovery like job creation, psychological recovery and social recovery. Using a gender perspective on the social policy response to the urgent needs of Tsunami-affected families and children, job creation for women survivors needs to be considered. In addition, mobilization of collective resources is required to develop social services for survivor families in need of urgent care.

THE ROLE OF SOCIAL WORKERS

Before the Tsunami, Thai social workers had focused their work mostly on social service delivery and social development. Their clients faced chronic problems like poverty. Now, their role is to assist in responding to and coping with crisis and crisis intervention. Resources provided for social service delivery in the past mostly were from the Thai government, private non-governmental organizations such as Association of Social Workers, National Council of Social Welfare (under the Royal Patronage) and local organizations. The Tsunami called for urgent assistance and collaboration from various sectors including local communities, regional, national and international organizations in both the short- and long-term. Their role is to assist in responding to and coping with crises brought about by the natural disaster.

According to Golan (1987), the basic theory and practice of crisis intervention is based on the following processes:

1. The increase of both internal and external stress throughout the life span that disrupts the state of equilibrium resulting from the surrounding environment
2. The impact of a disastrous event disturbs the individual balance and pushes the person into a vulnerable state with associated tension and anxiety
3. The person is in an "active crisis" state
4. The crisis is viewed as a threat to personal autonomy and well-being
5. Feelings of depression, mourning, and loss are experienced
6. The length of time between the beginning of the crisis stage and the final stage varies widely, but is usually no longer than six weeks
7. The dealing with possible loss of loved ones or family members
8. Reintegration is the ultimate goal for those who experience crisis events (Golan, 1987: pp. 364-365).

In light of this, Puryear (1988: pp. 50–51) describes the stages necessary in crisis intervention as follows:

Step 1: establishing communication and rapport
Step 2: assessing the problem
Step 3: assessing resources and strengths
Step 4: formulating a plan
Step 5: mobilizing the plan
Step 6: closing
Step 7: following up

To provide effective social services to Tsunami survivors, social workers in Thailand have to use both direct and indirect practices such as group work, family treatment, and expanding the knowledge base in social work practice. By applying Puryear's (1988) crisis intervention model, social workers have followed the process of crisis intervention. They have established communication and rapport, problem identification and assessment, mobilized resources and strengths, formulated a plan in a collaborative manner with local, national and international organizations and followed up. The intervention can be continued because of the establishment of temporary shelters by the Ministry of Social Development and Human Security.

Social workers have also developed educational programs for children such as a child protection system to provide knowledge for children in the southern areas. At the same time, Thailand needs to prevent the reoccurrence of a disaster such as that brought about by the Tsunami by establishing warning systems together with the development of educational systems for disaster prevention. As well as providing direct services to Tsunami survivors, social workers also needed to expand the knowledge base in social work practice such as working with foreigners who survived.

The Tsunami was a disaster and an unpredictable event which occurred without warning as no warning system was in place. Disasters may be defined as causing effects of severe stress on individuals, resulting in the loss of life and property (Krauss, 2005). Although efforts have been made to ameliorate the high levels of loss after the Tsunami, the needs have not been fully met, especially when examining the losses in the social context of the Tsunami. As stated by Krauss (2005), the waves took away a large number of people's lives, for example, in the village of Nam Khem. The survivors are still in the recovery process. As will be elaborated below, their needs have not been fully met with regard to both material and non-material assistance.

In Thailand, social workers have been working on disaster relief to address three levels of intervention–individual behavior, family behavior and community behavior. At the individual level, social workers have provided social assistance and counseling for individuals to cope with psychological stress after the Tsunami. The nature of their work requires a networking and interdisciplinary team approach including medical doctors and psychologists. At the family level, social services are provided for a large number of families who survived the Tsunami in Thailand. The ability to recover from the Tsunami trauma also depends on the family's ability to integrate and adapt, communication style, and level of mutual support. Family and children survivors of the Tsunami included foreigners and Thais, consisting of extended and nuclear families. Therefore, the crisis in this case may increase cohesiveness and strength or it may deplete family resources and lead to consequences such as desertion. Moreover, family and children survivors from various economic backgrounds ranging from elite tourists, who live in first-class hotels, to the poorest groups have different needs for reintegration into their societies. This has complicated the issues and areas of assistance in the long run. At the community level, it seems that the areas hit by the Tsunami required assistance across the broad spectrum of post-disaster needs such as shelter, clothing, cooking facilities, and communication and coordination systems. Also needed was the development of community cohesiveness, trained professionals, volunteers and indigenous leadership (Siporin, 1987).

In Thailand, social workers needed to work with professions from various backgrounds, national and international organizations, in response to clients' urgent needs. Further involvement in long-term prevention and curriculum development for disaster prevention and recovery from traumatic experiences were new roles to play. Examples of such plans are an educational scheme for the public and the survivors which serves to empower the target groups.

Case Study: Nam Khem Village

Nam Khem village is located in the southern part of Thailand. After the Tsunami hit in December 2004, a large number of survivors in need of help turned to the local hospitals. There were limited pharmaceutical and medical supplies. Assistance obtained from hospitals in Surat Thani included surgeons, nurses, and ambulances. There were both Thai and foreign patients (Krauss, 2005).

In the case of Nam Khem Village, the recovery process involves psychological supports as shown in the case of Puek and Lek below.

Puek and Lek are Tsunami survivors. Lek's father passed away because of the Tsunami waves. Puek heard the dreadful sound of the Tsunami wave. The noise of collapsing buildings and roaring water were so loud that Puek ran away with Lek. Puek realized later that what he heard was his imagination. They both obtained material assistance from the government camp including rice cookers, hotpots and radios. Wichien and Nang are also survivors. Wichien was in the hospital for about two weeks and obtained financial assistance of 9,000 Baht from the Tsunami fund for the victims provided by the Princess (Krauss, 2005).

The survivor cases of Lek, Puek, Wichien and Nang from Banh Nam Khem reflect the urgent need for responses including psychological support, financial and material assistance. In fact, Buddhist monks as well as social workers played a role in providing psychological support for the survivors.

Immediate contributions from UNICEF, UNDP, National Association of Social Workers (NASW), and the Ministry of Social Development and Human Security helped to alleviate and cope with the tragedy. Levels of assistance ranged from local supports by government and non-government organizations, to national and international organizations.

Thai Government

The Ministry of Social Development and Human Security provided services such as short-term occupational development, temporary shelters (soon-pakping-chaokrao) and social services like home visitations, occupational rehabilitation, psychological rehabilitation, family assistance, assistance provided for orphans and child care centers.

UNICEF, UNDP and some other international organizations considered the needs of foreign survivors. The Mental Health Department of Thailand provided psychiatric clinics in Kao Lak. Generally, the roles of social workers were those of crisis intervention, short- and long-term support.

Other social services provided by the Department of Health Services Supports included funds for training the unemployed and job creation. One of the actions taken was to revive Thailand's tourism industry in the Andaman provinces.

International Organizations

The international organizations that assisted in the relief and rehabilitation included UNICEF and UNDP.

Role of UNICEF

Children survivors of the Tsunami faced psychological problems such as anxiety, recurring sadness and nightmares, and traumatic experiences resulting from losing families, friends and homes. These children found help by attending counseling sessions so that they could talk about their experiences in a supportive way.

UNICEF's top priority is psychological care and support for children. It aims at supporting the work of psychological counselors and to heal and restore children with psychological activities in their communities and main social institutions like family and schools. For instance, children were asked to draw a picture of their home before and after the Tsunami. In this way, children were given a chance to express themselves.

At present, UNICEF currently works with the government and non-government agencies to support social protection and services based on educational and community services (UNICEF Thailand Newsletter, 2005).

Role of UNDP Thailand

As a long-term effort, UNDP supported the Royal Thai Government long-term recovery and rehabilitation effort to help communities such as Sea Gypsies and Muslim communities.

UNDP achievements and contributions were as follows (UNDP, 2005):

1. Longer-term community recovery programs have been formulated that will benefit 20,000 households in Tsunami-devastated areas.
2. Grassroots Credit schemes and revolving funds have been set up to finance the recouping of lost assets such as fishing boats, tackle and houses.
3. Alternative livelihood training and income generating activities such as batik, handicraft and woodworking are underway.
4. Activities to promote cultural heritage, environmental awareness and eco-tourism have been launched on Lanta Island.
5. Community groups have been formed in order to improve participation in local government decision-making and rehabilitation initiatives.

6. Coastal reforestation of mangrove and pine trees has been initiated by the communities to reduce the impact of a Tsunami in the future.
7. Multi-stakeholder consultation between local government, communities, NGOs, and the private sector is being organized.
8. Training selected government organizations in participatory planning to promote community participation is being initiated.
9. Networking and sharing of best practices between communities affected by the Tsunami are underway.
10. Close partnership was formed with the Department of Disaster Preparedness and Mitigation (DDPM) and the Asian Disaster Preparedness Centre (ADPC) to develop a project which complements other disaster management activities and is based on a comprehensive national risk management framework.
11. UNDP provides strategic support to the Royal Thai Government in the coordination of international technical assistance to the ongoing Tsunami relief and recovery efforts.
12. The End-to-End Early Warning System and Preparedness for Tsunami and Other Natural Hazards in Southern Thailand project seeks to develop an overall culture of safety that starts with a Tsunami early-warning system and ends with a heightened awareness of disaster planning and mitigation at all levels of Thai society.
13. Livelihood opportunities have been created, to assist in the repair of fishing boats, boat engines, nets and tackle for the affected communities.

RECOMMENDATIONS

Suggestions for Tsunami survivors' recovery process involve participation from local, national and international organizations in the following areas:

1. The use of educational technology to develop knowledge for international, national and local communities for disaster prevention. This should involve the establishment of short- and long-term courses in disaster prevention using the internet such as e-learning on disaster prevention.
2. Developing educational skills and computer technology for survivors and the public.

3. Identifying community resources and transferring the information using technology.
4. Coordinating with prominent associations, both international and national, such as UNICEF, National Social Welfare Council, National Social Workers Association of Thailand, World Buddhism Foundation, the Royal Institute of Thailand and educational institutions.
5. Developing an international network for disaster prevention focusing on the social work profession.
6. Seeking funds for long-term assistance for family and children survivors of the Tsunami.
7. Providing knowledge and critical skill development for social workers in order to cope with crises and work with clients from other cultural backgrounds and organizations.
8. Developing employment schemes such as area- and micro-based programs.

CONCLUDING REMARKS

The Tsunami recovery process has become a long-term challenge for the profession of social work in Thailand. To the casual eye, assistance work may only be focused on providing material aid. However, the survivors encountered social, economic and psychological impacts. It is hence necessary to employ a multidisciplinary perspective and to increase collaboration of social workers and other professions. With cooperative effort from local, national and international organizations, social workers have developed a healing process to aid the survivors of this crisis.

REFERENCES

Community Organization Development Institute (CODI) (2005). Tsunami-affected villages in Southern Thailand. www.codi.co.th/tsunami

Golan, N. (1987). Crisis intervention. *Encyclopedia of Social Workers*. Silver Spring, MD: National Association of Social Workers.

Krauss, E. (2005). *Wave of destruction: One Thai village and its battle with the Tsunami.* Chatham, Kent: Mackays of Chatham Ltd.

Manuta, J. B., Khrutmuang, S., & Lebel, L. (2005). The politics of recovery: Post-Asian Tsunami reconstruction in Southern Thailand. *Tropical Coasts*, 12(1), July 2005, 30-39.

Mass Communication of Thailand (MCOT) News (2005). *Tsunami*, 31 January 2005.
Puryear, D. A. (1988). *Helping people in crisis*. London: Jossey-Bass Publishers.
Siporin, M. (1987). Disasters and disaster aid. *Encyclopedia of Social Work*. Silver Spring, MD: National Association of Social Workers.
United Nations Development Programme (UNDP) (2005). *Report on the Tsunami recovery*. New York: United Nations.
United Nations International Children's Emergency Fund (UNICEF). *UNICEF Thailand Newsletter* (2005). Bangkok, Thailand: UNICEF.

doi:10.1300/J198v05n03_08

Index

Made in the USA
Monee, IL
13 September 2026

62977540R00095